Advances in the Processing and Application of Polymer and Its Composites II

Advances in the Processing and Application of Polymer and Its Composites II

Editors

Hui Zhao
Wei Wu
Bin Wang

MDPI

Basel • Beijing • Wuhan • Barcelona • Belgrade • Novi Sad • Cluj • Manchester

Editors

Hui Zhao
Guangxi University
Nanning
China

Wei Wu
Jihua Laboratory
Foshan
China

Bin Wang
Zhejiang Normal University
Jinhua
China

Editorial Office
MDPI
St. Alban-Anlage 66
4052 Basel, Switzerland

This is a reprint of articles from the Special Issue published online in the open access journal *Polymers* (ISSN 2073-4360) (available at: https://www.mdpi.com/journal/polymers/special_issues/ZA31J2PNML).

For citation purposes, cite each article independently as indicated on the article page online and as indicated below:

Lastname, A.A.; Lastname, B.B. Article Title. *Journal Name* **Year**, *Volume Number*, Page Range.

ISBN 978-3-7258-0337-8 (Hbk)
ISBN 978-3-7258-0338-5 (PDF)
doi.org/10.3390/books978-3-7258-0338-5

Contents

polymers

MDPI

Article

Self-Assembled Serpentine $Ni_3Si_2O_5(OH)_4$ Hybrid Sheets with Ammonium Polyphosphate for Fire Safety Enhancement of Polylactide Composites

Xiaohong Yi [1], Jingshu Huang [2], Yizhang Tong [2], Hui Zhao [3,*], Xianwu Cao [2] and Wei Wu [1,2,*]

1 Jihua Laboratory, Foshan 528200, China
2 Key Laboratory of Polymer Processing Engineering of Ministry of Education, South China University of Technology, Guangzhou 510640, China
3 Guangxi Key Laboratory of Calcium Carbonate Resources Comprehensive Utilization, College of Materials and Chemical Engineering, Hezhou University, Hezhou 542899, China
* Correspondence: zhh@gxu.edu.cn (H.Z.); wuwei@jihualab.ac.cn (W.W.)

Abstract: Biodegradable polylactide (PLA) has been widely utilized in people's daily lives. In order to improve the fire safety of PLA, ammonium polyphosphate (APP) was self-assembled onto the surface of serpentine $Ni_3Si_2O_5(OH)_4$ through the electrostatic method, followed by mixing with PLA by melt compounding. The APP-modified serpentine (serpentine@APP) dispersed uniformly in the PLA matrix. Compared with pure PLA, the PLA composite with 2 wt% serpentine@APP reduced the peak heat release rate (pHRR) and total heat release (THR) by 43.9% and 16.3%, respectively. The combination of APP and serpentine exhibited suitable synergistic flame-retardant effects on the fire safety enhancement of PLA. In addition, the dynamical rheological tests revealed that the presence of APP and serpentine could reduce the viscosity of PLA composites. The plasticizing effects of APP and serpentine benefited the processing of PLA. The mechanical properties of PLA/serpentine@APP maintained suitable performance as pure PLA. This study provided a feasible way to enhance the fire safety of PLA without sacrificing its mechanical properties.

Keywords: polylactide; serpentine; ammonium polyphosphate; composite; flame retardant; mechanical property

Citation: Yi, X.; Huang, J.; Tong, Y.; Zhao, H.; Cao, X.; Wu, W. Self-Assembled Serpentine $Ni_3Si_2O_5(OH)_4$ Hybrid Sheets with Ammonium Polyphosphate for Fire Safety Enhancement of Polylactide Composites. *Polymers* **2022**, *14*, 5255. https://doi.org/10.3390/polym14235255

Academic Editor: Ali Reza Zanjanijam

Received: 9 November 2022
Accepted: 30 November 2022
Published: 1 December 2022

1. Introduction

Biodegradable polymers have drawn considerable attention in both academia and industry due to the increasing awareness of environmental pollution problems. Polylactide (PLA) is a kind of bio-based and biodegradable polyester that shows promising applications in various fields, such as packaging, automotives, and electronics [1–4]. It exhibits unique properties such as suitable transparency, high mechanical strength, suitable processibility, and non-toxicity. However, the intrinsic flammability of PLA limits its further applications and development. Therefore, it is an important and urgent issue to improve the fire safety performance of PLA [5–7]. Recently, flame-retardant PLA composites containing different types of fillers, such as montmorillonite [8], carbon black [9], sepiolite [10], and carbon nanotubes [11], have been extensively investigated. The addition of organic modified montmorillonite (OMMT) enhanced the thermal stability and fire resistance of PLA significantly due to its excellent physical barrier effect and char-forming effect [8]. Wen et al. [9] grafted DOPO onto the surface of carbon black (CB-g-DOPO). The results revealed that CB-g-DOPO was effective at improving the flame retardancy and mechanical properties of PLA/CB composites. The peak heat release rate (pHRR) of the PLA composite containing 10 wt% CB-g-DOPO was reduced by 40.7% as compared with that of pure PLA. Jiang et al. decorated natural sepiolite with DOPO (SEP-g-DOPO) through the reaction between amino groups and salicylaldehyde. The presence of SEP-g-DOPO benefited the generation of the dense and continuous char layer, protecting the interior layer more efficiently durin combustion [10].

Two-dimensional (2D) layered materials have exhibited superior flame-retardant be haviors in a variety of polymer composites due to their barrier effect and promotion o char residues [12,13]. Many researchers have explored 2D-based nanofillers in combi nation with flame-retardant additives to improve the fire performance of PLA [14]. A combination of graphitic carbon nitride (g-C_3N_4) with melamine pyrophosphate (MPP) o DOPO exhibited a significant reduction in pHRR and total heat release (THR) of PLA [15 Xu et al. utilized melamine-cyanuric acid (MCA) hybrids as a shell layer to decorate lamel lar molybdenum disulfide (MoS_2) plates and then mixed them with PLA through mel blending [16]. The core-shell structure of MoS_2 could suppress the pyrolysis rate anc promote the graphitization degree of char residues. Jing et al. demonstrated that the core-shell flame-retardant/graphene oxide (GOH) hybrids could reduce fire hazards anc improve the toughness of PLA simultaneously [17]. Zhang et al. reconstructed layerec double hydroxides (LDH) with phosphotungstic acid [18]. The PLA composite obtained the UL-94 V-0 rating and achieved a maximal LOI value of 48.3% with 18.0 wt% intumescen flame retardant (IFR) and 2.0 wt% modified LDH. The modified LDH exhibited suitabl synergistic effects with intumescent flame retardant on the enhancement of fire resistanc of PLA.

Serpentine [$Ni_3Si_2O_5(OH)_4$] is a type of transition metal silicate hydroxide that exists extensively in the oceanic lithosphere [19]. It is a silicate with unlimited two-dimensiona extension made of Si-O tetrahedrons. The oxygen atop the unshared siloxane tetrahedron possesses a residual negative charge (Scheme 1), which can react with Ni^{2+} to produce Ni-O octahedron sheets. Lamellar serpentine nanosheets have been widely utilized fo catalysis [20], energy storage [21], and sensors [22]. It is rich in OH groups and has a similar structure to LDH, making it potentially useful in the flame-retardant sector of polyme matrices. However, the application of serpentine as a promising flame retardant in polyme composites is rarely reported.

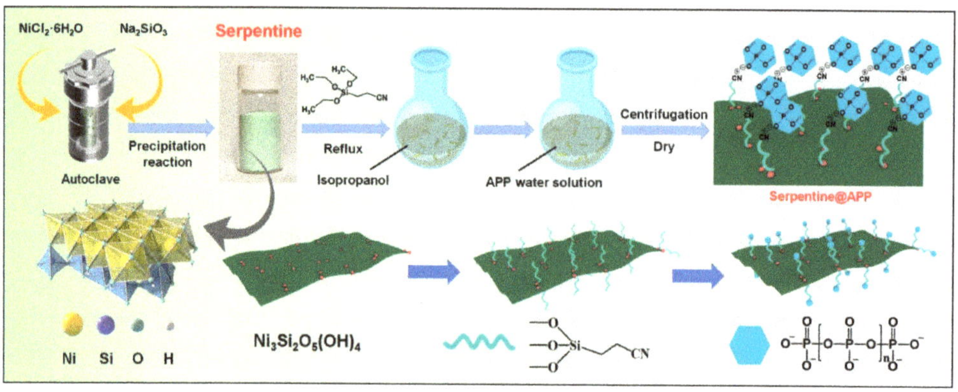

Scheme 1. The synthesis route of serpentine@APP.

In this work, lamellar serpentine nanosheets were synthesized by precipitation reactions under hydrothermal conditions. Then ammonium polyphosphate (APP) was grafted onto the serpentine nanosheets via electrostatic self-assembly by using 2-cyanoethyltriethoxysilane as a chemical bridge. The APP-functionalized serpentine (serpentine@APP) was incorporated into the PLA matrix by melt compounding as a novel type of flame-retardant hybrid filler. The effects of the as-fabricated serpentine@APP on the rheological properties, mechanical properties, and flame-retardant performance of PLA composites were also investigated.

2. Experimental

2.1. Materials

PLA (Ingeo 4032D) with a density of 1.25 g/cm^3 was supplied by Natureworks (Plymouth, MN, USA). Nickel chloride hexahydrate and sodium silicate were purchased from Aladdin Bio-Chem Co., Ltd. (Shanghai, China). 2-cyanoethyltriethoxysilane, ammonium polyphosphate (APP, n < 20), and isopropanol were supplied by J&K Co., Ltd. (Beijing, China).

2.2. Preparation of Serpentine@APP Hybrid

The synthesis route of ammonium polyphosphate decorated serpentine (serpentine@APP) was shown in Scheme 1. The serpentine $Ni_3Si_2O_5(OH)_4$ nanosheets were synthesized by precipitation reactions under hydrothermal conditions [23]. Typically, 1 mmol of Na_2SiO_3 and 35 mL of DI water were mixed by vigorous stirring. Then, 1.5 mmol of $NiCl_2$ was introduced to the Na_2SiO_3 solution to generate light green precipitates. Subsequently, the as-prepared suspension was moved into a 100 mL Teflon-lined stainless steel container and heated in an oven at 180 °C for 12 h. To remove the excess OH^-, Cl^-, and Na^+ ions, the resulting hydrothermal treatment materials were centrifuged and rinsed with DI water and isopropanol, respectively. The washed precipitate was collected after being dried at 120 °C.

1.0 g of as-prepared $Ni_3Si_2O_5(OH)_4$ nanosheets were mixed with 200 mL of isopropanol by ultrasonication. Then the solution was treated with 100 µL of 2-cyanoethyltriethoxysilane dropwise for 2 h at 90 °C, followed by centrifugation and rinsing with DI water and isopropanol, respectively. A total of 5.0 g of APP was dissolved in 200 mL of DI water with vigorous stirring to form a clear solution. After the APP was dispersed completely, 1.0 g of silane-modified serpentine was subsequently added with vigorous stirring. Then the mixture was centrifuged at 10,000 rpm for 15 min and dried in a vacuum oven. The obtained APP-functionalized serpentine was denoted as serpentine@APP.

2.3. Preparation of PLA Composites

A counter-rotating internal mixer (PLASTIC-ORDER, Brabender, Germany) with a rotation speed of 60 rpm was used to melt compound PLA composites containing 2 wt% different types of fillers for 10 min at 180 °C. Following that, the samples were hot-molded (185 °C, 10 min) into specimens of various sizes. The formulations of PLA composites are shown in Table 1.

Table 1. Formulations of pure PLA and its composites.

Sample	PLA (wt%)	APP (wt%)	Serpentine (wt%)	Serpentine@APP (wt%)
PLA	100	–	–	–
PLA/APP	98	2	–	–
PLA/serpentine	98	–	2	–
PLA/serpentine@APP	98	–	–	2

2.4. Characterization

The morphology of serpentine, serpentine@APP, and PLA composites was observed by scanning electron microscopy (SEM, FEI, Quata 250, Hillsboro, FL, USA). Before the SEM observations, the samples were sputtered with a thin layer of gold. An atomic force microscope (AFM, VEECO Nanoscope, Plainview, NY, USA) was used to measure the thickness and morphology of the synthesized serpentine nanosheets in tapping mode. The Fourier transform infrared (FT-IR) spectra were carried out on a Fourier transform infrared spectrometer (Nicolet, Nexus 670, Ramsey, MN, USA) in the range of 400–4000 cm^{-1}. The elemental compositions were analyzed by X-ray photoelectron spectroscopy (XPS, Physical Electronics, PHI 5802 spectrometer, Chanhassen, MN, USA). The thermal stability was investigated by a thermogravimetric analyzer (TGA, Netzsch, TGA-209F3, Selb, Germany) in the temperature range of 30–700 °C under nitrogen flow. The heating rate was fixed at 10 °C/min. The limiting oxygen index (LOI) was measured on an HC-2 oxygen index instrument (Jiangning Analytical Instrument, Jiangning, China). The UL-94 rating of the samples was evaluated by a vertical burning tester (FTT, Derby, UK). The fire characteristics of the PLA composites were measured on a cone calorimeter (FTT, UK) under an external heat flux of 35 kW/m^2 based on the standard of ISO-5660. The structure of char residues was analyzed by a Raman spectrometer (LabRAM ARAMIS, Villeneuve d'Ascq, Paris, France). The dynamical rheological properties were measured on an Anton-Paar MCR-302 dynamic rheometer (Graz, Austria). A dynamic mechanical analyzer (Netzsch, DMA 242, Weimar, Germany) was utilized to examine the dynamic mechanical properties in tensile mode. The rectangular specimens with a dimension of 30 × 4 × 0.5 mm^3 were measured in the temperature range of 0–80 °C at a fixed heating rate of 3 °C/min. The tensile properties were evaluated on an electronic testing machine (Type 5566, Instron, Norwood, MA, USA). The dumbbell specimens with 75 × 4 × 1 mm^3 were tested at a fixed speed of 1 mm/min.

3. Results and Discussion

3.1. Characterization of Serpentine@APP

The morphology of the prepared serpentine is shown in Figure 1a. The serpentine nanosheets exhibit a layered structure with a smooth surface. The AFM image in Figure 1b confirms the lamellar structure. The height of the region marked by the white line in the AFM image ranges from 0.6 to 0.7 nm, corresponding to the height of monolayer serpentine As for serpentine@APP, a rougher surface of serpentine@APP is clearly observed in Figure 1c,d, which is due to the charge self-assembling between APP and the silane-modified serpentine. In Figure 1e, the corresponding EDX mapping images confirm the existence of Si, Ni, P, and N elements, which disperses evenly onto the surface of serpentine@APP, indicating APP has been successfully decorated onto the surface of serpentine.

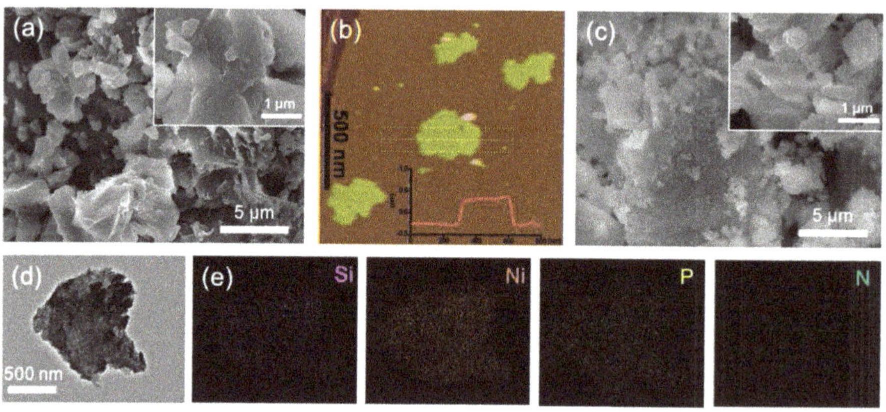

Figure 1. (**a**) SEM images of serpentine, (**b**) AFM image of serpentine (Insert is the height of the selected curve); (**c**) SEM image, (**d**) TEM image, and (**e**) the corresponding EDX elemental mapping of Si, Ni, N, and P elements of serpentine@APP.

The XRD patterns of APP, serpentine, and serpentine@APP are depicted in Figure 2a. The four diffraction peaks at 20.4°, 25.3°, 34.1°, and 35.3° are assigned to the (110), (004), (200), and (202) planes of serpentine, respectively, which are consistent with the serpentine card (JCPDS no. 49-1859) [24,25]. Moreover, the diffraction peaks of serpentine are very broad, which is consistent with the nanostructured nature of this material. The XRD patterns indicate that the $Ni_3Si_2O_5(OH)_4$ nanosheets were synthesized successfully. There are several small diffraction peaks between 20° and 31° for serpentine@APP, which are attributed to the grafted APP. In addition, the diffraction peak at 32.7° in APP disappears in the pattern of serpentine@APP, suggesting there is a chemical interaction between APP and serpentine. The FT-IR spectra of serpentine and serpentine@APP are shown in Figure 2b. The sharp peaks at 3413 cm^{-1} and 1618 cm^{-1} are generated by the -OH vibration in $Ni_3Si_2O_5(OH)_4$. The characteristic peak at 1042 cm^{-1} is owing to the Ni–O–Si chemical bond [26]. The band at 619 cm^{-1} corresponds to the Ni–O bond. In addition, the peak at 484 cm^{-1} is attributed to the symmetric stretching vibration of the Si–O bond [27]. These characteristic peaks further prove the formation of nickel silicate in the hybrid. The chemical compositions of serpentine and serpentine@APP were evaluated by XPS, and the corresponding data are shown in Figure 2c–e. In Figure 2c, the presence of Ni, Si, C, and O is clearly observed in the full scan of the serpentine. The appearance of C may be attributed to pollution from the environment. The high-resolution spectra of N 1s and P 2p of serpentine@APP in Figure 2d,e confirm that the APP was attached to the serpentine successfully. To measure the thermal stability of APP, serpentine, and serpentine@APP under a nitrogen atmosphere, the TGA curves are shown in Figure 2f, and the corresponding TGA data are shown in Table S1. The pure APP begins to degrade (T_{10}, the temperature that corresponds to 10% weight loss) at 309.4 °C and to decompose most rapidly (T_{max}, the temperature that corresponds to the maximum weight loss rate) at 357.1 °C. As for serpentine and serpentine@APP, both of them exhibit a quick decomposition stage below 100 °C, which is attributed to the loss of absorbed water. With the decoration of APP, the T_{10} for serpentine@APP increases from 106.5 to 182.1 °C as compared with pure serpentine. It is because the decorated APP may serve as a protective coating, delaying the decomposition of the serpentine [28]. These above results indicated that serpentine@APP had better thermal stability than pure serpentine.

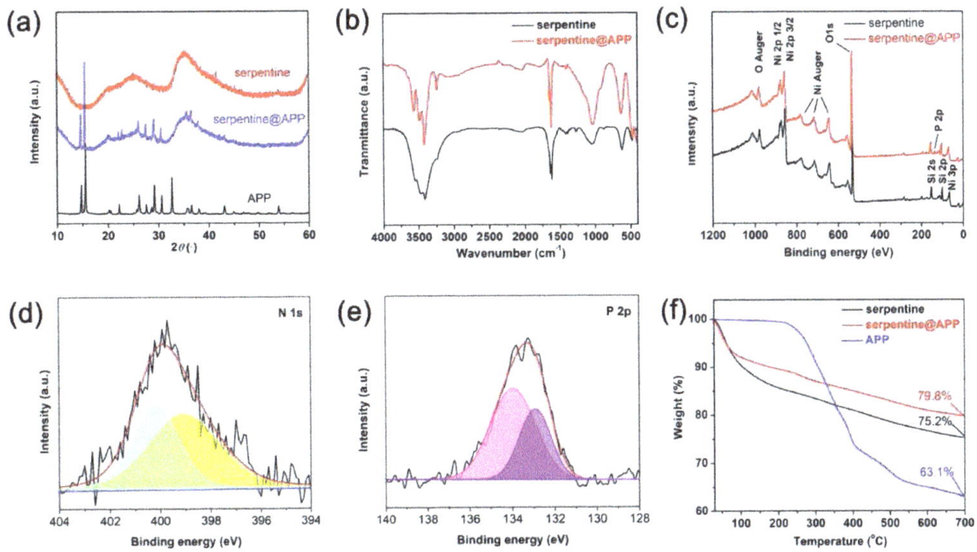

Figure 2. (**a**) XRD patterns (**b**) FT-IR spectra (**c**) XPS full scan spectra of serpentine and serpentine@APP; (**d**) N 1s, (**e**) P 2p high-resolution spectra of serpentine@APP, and (**f**) TGA curves of APP, serpentine, and serpentine@APP.

3.2. Fractured Surface Analysis of PLA Composites

The morphology of the cryo-fractured surfaces of the pure PLA and its composites is shown in Figure 3. It is evident that pure PLA has a smooth fractured surface (Figure 3a) due to its brittle fracture at low temperatures. In Figure 3b, some APP particles are exposed on the smooth surface of the PLA/APP composite. With the incorporation of serpentine, the PLA/serpentine composite (Figure 3c) exhibits a much rougher surface than that o pure PLA. That is because the presence of rigid serpentine fillers can absorb more energy and inhibit crack propagation during fracture. As for the PLA/serpentine@APP composite (Figure 3d), the introduction of serpentine@APP also increases the cracks in the surface. In addition, there are no cavities on the surfaces of PLA/serpentine and PLA/serpentine@APP composites, indicating that both serpentine and serpentine@APP have suitable interfacia interaction with the PLA matrix.

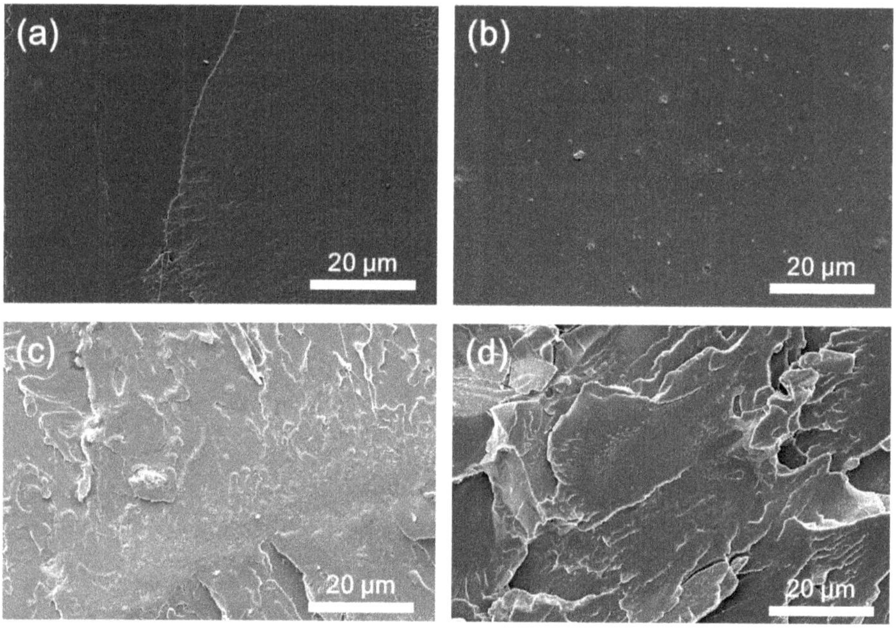

Figure 3. SEM images of the cryo-fracture surfaces of (**a**) pure PLA, (**b**) PLA/APP, (**c**) PLA/serpentine and (**d**) PLA/serpentine@APP composites.

3.3. Thermal Stability

The thermal stability of PLA composites was evaluated by TGA, as shown in Figure 4. The temperatures at 10% weight loss (T_{10}), the temperatures at the maximum weight loss rate (T_{max}), and the char residues at 800 °C are listed in Table 2. All of the samples in Figure 4a show a single primary decomposition stage between 300 and 400 °C, which corresponds to random chain scission and particular chain-end scission. [29]. With the incorporation of APP, the T_{10} of the PLA/APP composite increases from 346.2 to 349.7 °C as compared with pure PLA. However, the incorporation of serpentine results in a slight decrease in T_{10}. It is speculated that serpentine containing a large number of hydrogen groups has lower thermal stability, as shown in Figure 2f. After decoration with APP, the T_{10} of the PLA/serpentine@APP shifts to a higher value (346.0 °C). The T_{max} for pure PLA, PLA/APP, PLA/serpentine, and PLA/serpentine@APP in Figure 4b is 379.0, 377.1, 373.5, and 376.7 °C, respectively. As shown in Table 2, almost no char residue is observed for the pure PLA at 800 °C. Meanwhile, the presence of APP and serpentine can increase the char residue of PLA composites. The increased char residue weight can contribute to the enhancement of fire safety for PLA composites.

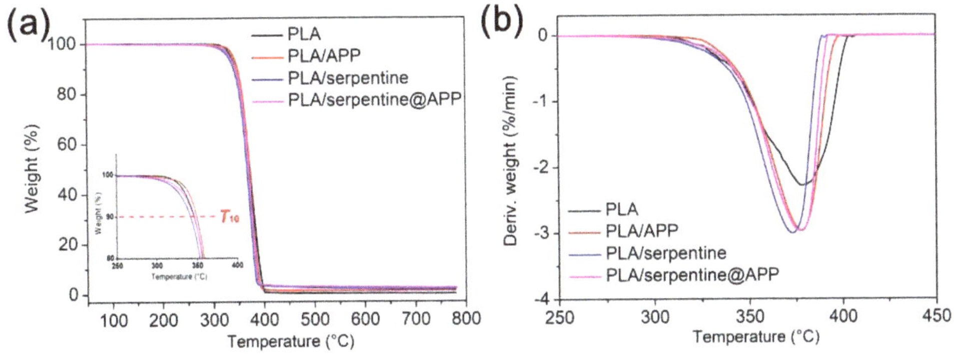

Figure 4. Thermal stability of pure PLA and its composites, (**a**) TGA, (**b**) DTG.

Table 2. TGA data for pure PLA and its composites.

Samples	T_{10} (°C)	T_{max} (°C)	Residues at 800 °C (wt%)
PLA	346.2	379.0	0.21
PLA/APP	349.7	377.1	1.39
PLA/serpentine	341.5	373.5	1.72
PLA/serpentine@APP	346.0	376.7	1.95

3.4. Flame-Retardant Performance

The limited oxygen index (LOI) and UL-94 test results are summarized in Table 2. It is clear that pure PLA has an LOI value of 19 and no rating in the UL-94 test, indicating it is a flammable material. With the addition of serpentine@APP, the LOI value of PLA composite increases from 19 to 25.5 but can only achieve a UL-94 V-1 rating. However, all of the samples have dripping phenomena due to their low melting viscosity.

A cone calorimeter is a useful tool to evaluate the fire safety of polymers in real fire accidents [30]. The heat release rate (HRR), total heat release (THR), and char residues of PLA composites as a function of time are presented in Figure 5, and more detailed parameters, including time to ignition (TTI), peak heat release rate (pHRR) are summarized in Table 3. In Figure 5a, pure PLA exhibits a pHRR value of 507.8 kW/m^2 and a THR value of 57.1 MJ/m^2. With the addition of APP, the pHRR of PLA/APP decreases to 419.7 kW/m^2, which is decreased by 17.3% as compared to pure PLA. This is because the thermal decomposition of APP can release free radical scavengers (·P and ·PO), which may trap flammable radicals (·O, ·H, and ·OH), inhibit the burning reaction chain, and generate more solid products [31]. As for PLA/serpentine, the presence of serpentine can serve as a thermal barrier to prolong the escape paths for the flammable gases. In addition, the nickel element in serpentine has a suitable catalytic carbonization effect that promotes the formation of the char layer [32]. The PLA/serpentine@APP exhibits a pHRR value of 284.7 kW/m^2 and a THR value of 47.8 MJ/m^2, which are reduced by 43.9% and 16.3%, respectively, as compared to those of pure PLA. In Figure 5b, it is clear that the slope of PLA/serpentine@APP is lower than those of other curves, indicating that the combination of APP and serpentine has suitable synergistic flame-retardant effects that suppress heat release during combustion.

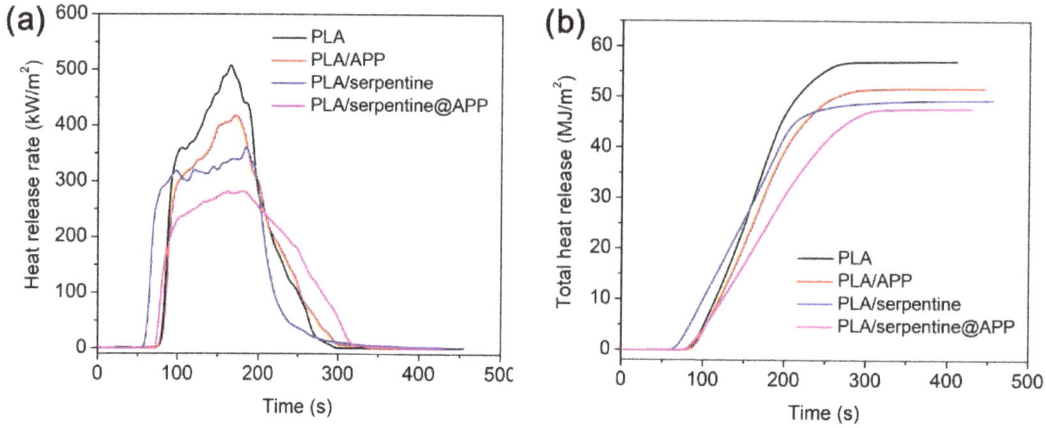

Figure 5. Cone calorimeter tests of pure PLA and its composites. (**a**) HRR, (**b**) THR.

Table 3. LOI, UL-94 test, and cone calorimeter data for pure PLA and its composites.

Samples	LOI (%)	UL-94 Rating	Dripping	TTI (s)	PHRR (kW/m^2)	THR (MJ/m^2)
PLA	19	NR	Yes	75	507.8	57.1
PLA/APP	22	V-2	Yes	77	419.7	51.8
PLA/serpentine	23.5	V-2	Yes	68	362.4	49.5
PLA/serpentine@APP	25.5	V-1	Yes	72	284.7	47.8

3.5. Char Residues Analysis

Figure 6 shows the photographs of char residues of PLA composites after the cone calorimeter test. It is noted that pure PLA has been completely burned, and very little char residue can be found in the aluminum foil. Meanwhile, the char residue of PLA/serpentine shows a fluffy and discontinuous state. With the addition of APP, an extremely thin and dense carbon layer is formed on the sample of PLA/APP after burning and adhering to the tin foil at the bottom. Moreover, the char residue of PLA/serpentine@APP is relatively continuous, and a large carbon block is formed on the surface. To investigate the microstructure of char residues, the SEM images of the char residues are shown in Figure 7. PLA/APP has a compact and continuous char layer in Figure 7a due to the presence of phosphorus derivatives in APP. It is observed that the lamellarly structured serpentine of PLA/serpentine is loosely distributed after combustion, and the area of the continuous carbon layer is small (Figure 7b). While the serpentine in the char residue of PLA/serpentine@APP is bonded together to form an effective carbon layer, which acts as a barrier to the diffusion of O_2 and heat, preventing further combustion of the substrate. A small amount of pores appears in the carbon layer of PLA/serpentine@APP (Figure 7c), which may be caused by the release of incombustible gases (CO, CO_2, and NH_3) into the gas phase by APP before the decomposition of the PLA matrix during the combustion process [33]. The presence of serpentine@APP contributes to the formation of more dense and complete carbon layers, which can more effectively prevent the exchange of heat and pyrolysis gas products with the external area [34], thereby delaying and restricting the development of combustion to achieve the flame-retardant effect.

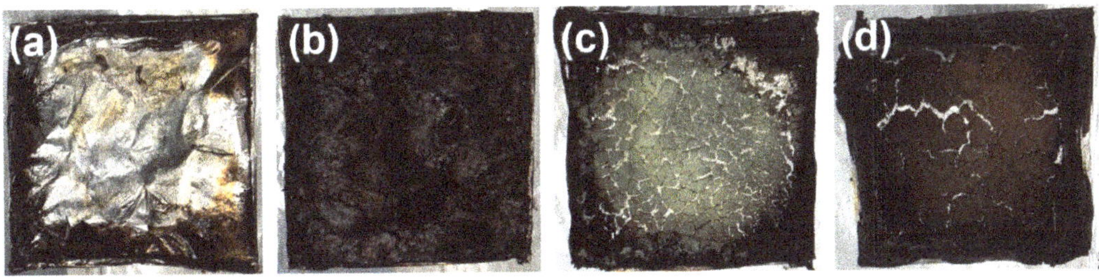

Figure 6. Digital images of char residues of the PLA samples after cone tests. (**a**) pure PLA, (**b**) PLA/APP, (**c**) PLA/serpentine, and (**d**) PLA/serpentine@APP.

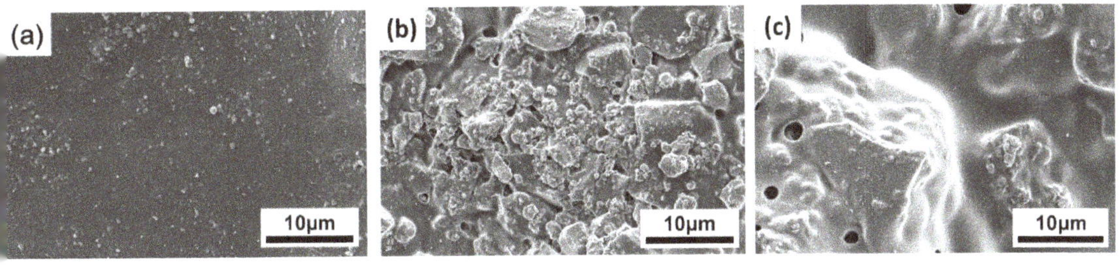

Figure 7. SEM images of the char residues of (**a**) PLA/APP, (**b**) PLA/serpentine, and (**c**) PLA/serpentine@APP.

The chemical composition of the char residues after the cone calorimeter test was analyzed by XPS, as shown in Figure 8. The XPS full scan spectra in Figure 8a confirm the existence of C, O, Ni, P, and Si elements in the char residues. In Figure 8b,c, the high resolution of the C 1s XPS spectra reveals that the peaks of the C=O (290.7 eV), C–O (288.3 eV), and C–C (284.8 eV) bonds can be fitted separately. Furthermore, the content of C=O and C–O bonds in the carbon residue of the PLA/serpentine sample is lower than that of the PLA/serpentine@APP sample. The molten flow of the PLA matrix during combustion breaks up the carbon layer formation, leading to more contact with air and promoting a more complete combustion of PLA/serpentine. However, it is difficult for the APP to flow with the PLA matrix melt in the PLA/serpentine@APP sample. Meanwhile, it will accelerate the dehydration of the matrix, which is conducive to the rapid formation of a preliminary complete carbon layer after combustion, thereby reducing the contact with O_2, promoting the incomplete combustion of the matrix, and increasing the content of C=O and C–O [35]. The high resolution of the P 2p spectrum in Figure 8d further confirms the presence of phosphorus groups in PLA/serpentine@APP. The peaks of the P–O (133.7 eV) and P=O bond (132.4 eV) can be fitted separately [36]. Previous studies have demonstrated that the accumulation of phosphorus-containing compounds on the surface of the carbon layer will reduce the permeability of the carbon layer, which is more conducive to improving the barrier properties of the carbon layer [37].

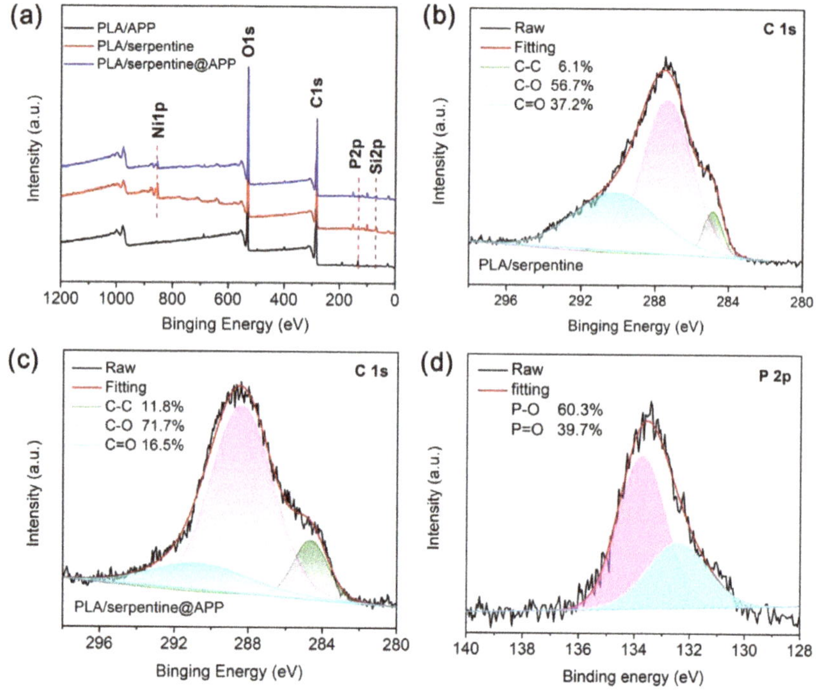

Figure 8. XPS spectra of the char residues after cone calorimeter test. (**a**) Full scanning spectra, (**b**) C 1s spectrum of PLA/serpentine; (**c**) C 1s, and (**d**) P 2p spectra of PLA/serpentine@APP.

The graphitization degree of the char residues was analyzed by Raman spectra [38]. The two broad bands around 1380 and 1590 cm^{-1} in Figure 9 correspond to the disordered carbon structure of the D band and the graphitized structure of the G band, respectively [39,40]. The integral strength ratio (I_D/I_G) of the D band and the G band can measure the degree of graphitization. The lower the ratio, the higher the degree of graphitization of the carbon layer [41,42]. It is noted that the I_D/I_G of char residues of PLA/serpentine@APP is 3.00 (Figure 9a), which is lower than those of PLA/serpentine (I_D/I_G = 3.67, Figure 9b) and PLA/APP (I_D/I_G = 3.39, Figure 9c). The results show that the APP-modified serpentine can obtain a carbon layer with a higher degree of graphitization than the serpentine alone. APP will form pyrophosphoric acid or polyphosphoric acid in the process of thermal decomposition [43], which will destroy the PLA bone chain and the hydroxyl group in the matrix to form a higher-quality carbon layer. The highly graphitized carbon structure layer can act as a flame-retardant barrier that keeps the fire from spreading during the combustion process of the substrate.

3.6. Rheological Behavior of PLA Composites

The linear viscoelastic region of pure PLA and its composites was determined by the dynamical strain sweep tests [44,45], as depicted in Figure S1. All of the materials show a linear viscoelastic domain up to 30% strain. The apparent shear-thinning phenomena occur when the shear amplitude is above 30% strain. Therefore, the dynamical frequency sweep experiment was carried out using a strain amplitude of 1% to avoid non-linear behavior at low frequencies [46]. The complex viscosity (η*), storage modulus (G'), loss modulus (G''), and damping factor (tanδ) as a function of frequency are shown in Figure 10. It is noted that a Newtonian plateau in the low-frequency region is observed for all of the samples in Figure 10a, followed by a shear-thinning tendency at the high-frequency region. In addition, the shear-thinning phenomenon of pure PLA is more profound than that of

PLA/sepentine@APP, indicating that the presence of sepentine@APP contributes to the disentanglement of PLA chains. Interestingly, it is noted that the addition of serpentine will result in a decrease in the η^* of PLA composites. It can be attributed to the plasticizing effect of the 2D serpentine, which can decrease the rectangle of the polymer chains. In Figure 10b,c, the G' and G'' show a similar decrease tendency. However, the PLA composite containing serpentine has a greater storage modulus at low frequencies and a more pronounced decline at high frequencies. This phenomenon may be attributed to the 2D serpentine flake being oriented at a high rotation speed, which is consistent with other 2D materials. Interestingly, the PLA/sepentine@APP composite has a much lower storage modulus than that of APP or serpentine applied solely in the high-frequency region. It is speculated that the low molecular weight APP has a suitable synergistic plasticizing effect with serpentine. The value of tanδ represents the ratio between G' and G'', which indicates the dissipation of heat energy. In Figure 10d, the values of tanδ are decreased obviously with the addition of serpentine and sepentine@APP at the low-frequency region. This is because the presence of serpentine or sepentine@APP contributes to the elasticity improvement of PLA.

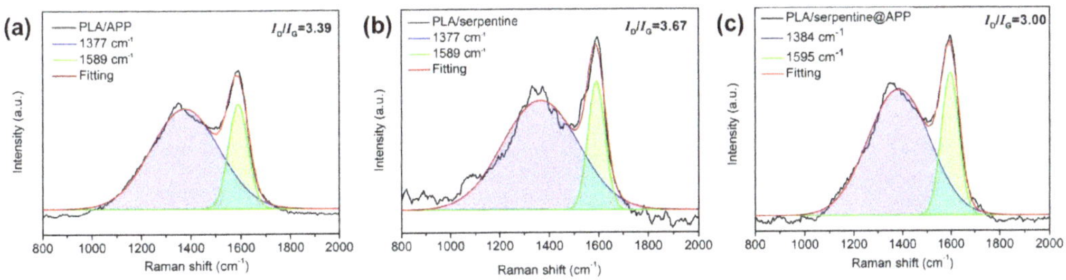

Figure 9. Raman spectra of the char residues of (**a**) PLA/APP, (**b**) PLA/serpentine, and (**c**) PLA/serpentine@APP.

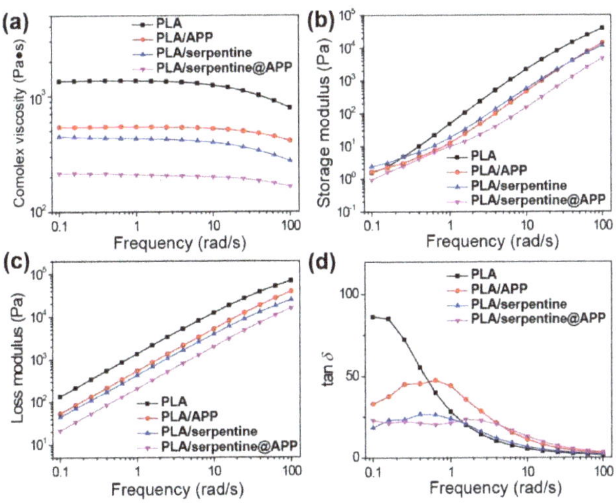

Figure 10. Dynamical frequency sweep of PLA composites. (**a**) complex viscosity, (**b**) storage modulus, (**c**) loss modulus, and (**d**) loss factor.

3.7. Mechanical Properties of PLA Composites

Figure 11a displays the tensile stress versus strain curves of PLA composites, and the relevant data are provided in Table 4. In Figure 11a, all of the samples display the typical brittle behavior that breaks at certain places quickly after the yielding point. The pure PLA has a maximum tensile stress of 59.4 MPa, and an elongation at break of 6.05%. The stiffness of serpentine is considered to be the reason why the Young's modulus of PLA/serpentine jumps from 1476.6 to 1553.3 MPa when compared to pure PLA. However the mechanical properties of PLA/serpentine@APP somewhat deteriorate when compared to PLA/serpentine. That is because the APP has a small molecular weight and is grafted onto the surface of the serpentine by electrostatic interaction, leading to reduced tensile strength and elongation at break. As shown in Figure 11b, the storage modulus of PLA/APP is lower than that of pure PLA in the whole temperature range. Accordingly, it can be explained by the fact that the addition of APP has a plasticizing effect on PLA.

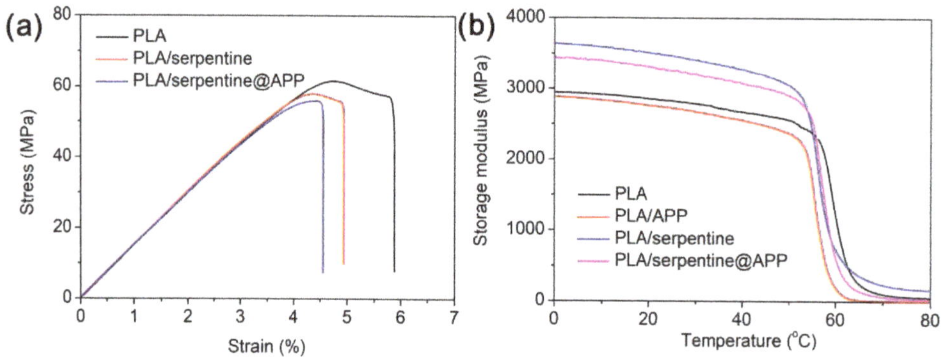

Figure 11. (**a**) Tensile stress versus strain of tensile test, (**b**) storage modulus of dynamical mechanical analysis of pure PLA and its composites.

Table 4. Tensile properties for pure PLA and its composites.

Samples	Tensile Stress (MPa)	Young's Modulus (MPa)	Elongation at Break (%)
PLA	59.4 ± 2.3	1476.6 ± 74.1	6.05 ± 0.49
PLA/serpentine	57.9 ± 2.8	1553.3 ± 73.9	4.97 ± 0.56
PLA/serpentine@APP	56.6 ± 1.5	1493.2 ± 44.3	4.83 ± 0.21

4. Conclusions

In this study, serpentine $Ni_3Si_2O_5(OH)_4$ was first synthesized by precipitation reactions under hydrothermal conditions. Then APP was electrostatically grafted onto the surface of serpentine to create a novel type of hybrid flame-retardant additive (serpentine@APP), followed by mixing with PLA by melt compounding. The PLA composite containing 2 wt% serpentine@APP exhibited suitable fire safety enhancement with a 43.9% reduction in pHRR and a 16.3% reduction in THR as compared to those of pure PLA. This is because the decomposition of APP will release free radical scavengers, the thermal barrier effect of serpentine, and the catalytic carbonization effect of the nickel element in serpentine. The presence of serpentine@APP could reduce the complex viscosity of PLA, which improved its processability. In addition, the tensile stress and elongation at the break of PLA/serpentine@APP showed a small decrease as compared to pure PLA. This research offers a feasible strategy to develop a flame-retardant hybrid filler to increase the fire safety of PLA.

Supplementary Materials: The following supporting information can be downloaded at https://www.mdpi.com/article/10.3390/polym14235255/s1, Figure S1: Strain sweep of pure PLA and its composites; Table S1: TGA data of APP, serpentine and serpentine@APP.

Author Contributions: Conceptualization, W.W.; methodology, X.Y., J.H. and Y.T.; validation, X.Y. and J.H.; investigation, J.H. and Y.T.; resources, X.Y., W.W. and X.C.; data curation, H.Z.;

writing—original draft preparation, X.Y. and J.H.; writing—review and editing, H.Z. and W.W.; supervision, X.C. and W.W.; project administration, W.W. and X.Y.; funding acquisition, X.C., W.W. All authors have read and agreed to the published version of the manuscript.

Funding: This work was supported by the Guangdong Basic and Applied Basic Research Foundation (2021A1515012425), the International collaboration research fund of Guangdong Province (2020A0505100010), the Opening Project of Guangxi Key Laboratory of Calcium Carbonate Resources Comprehensive Utilization (HZXYKFKT202204), and the Industry Leading Talents Projects of Guangzhou (201809010011).

Institutional Review Board Statement: Not applicable.

Informed Consent Statement: Not applicable.

Data Availability Statement: The raw/processed data generated in this work are available upon request from the corresponding author.

Conflicts of Interest: The authors declare no conflict of interest.

References

1. Raquez, J.M.; Habibi, Y.; Murariu, M.; Dubois, P. Polylactide (PLA)-based nanocomposites. *Prog. Polym. Sci.* **2013**, *38*, 1504–1542. [CrossRef]
2. Armentano, I.; Bitinis, N.; Fortunati, E.; Mattioli, S.; Rescignano, N.; Verdejo, R.; Lopez-Manchado, M.A.; Kenny, J.M. Multifunctional nanostructured PLA materials for packaging and tissue engineering. *Prog. Polym. Sci.* **2013**, *38*, 1720–1747. [CrossRef]
3. Wu, W.; Han, S.; Venkatesh, S.; Sun, Q.; Peng, H.; Zhou, Y.; Yeung, C.; Li, R.K.Y.; Roy, V.A.L. Biodegradable skin-inspired nonvolatile resistive switching memory based on gold nanoparticles embedded alkali lignin. *Org. Electron.* **2018**, *59*, 382–388. [CrossRef]
4. Wu, W.; Cao, X.W.; Zhang, Y.J.; He, G.J. Polylactide/halloysite nanotube nanocomposites: Thermal, mechanical properties, and foam processing. *J. Appl. Polym. Sci.* **2013**, *130*, 443–452. [CrossRef]
5. Tawiah, B.; Yu, B.; Yuen, A.C.Y.; Yuen, R.K.K.; Xin, J.H.; Fei, B. Thermal, crystalline and mechanical properties of flame retarded poly(lactic acid) with a PBO-like small molecule-phenylphosphonic bis(2-aminobenzothiazole). *Polym. Degrad. Stab.* **2019**, *163*, 76–86. [CrossRef]
6. Tawiah, B.; Yu, B.; Yang, W.; Yuen, R.K.K.; Fei, B. Flame retardant poly(lactic acid) biocomposites based on azo-boron coupled 4,4′-sulfonyldiphenol and its combination with calcium lignosulfonate-crystalline and mechanical properties. *Polym. Adv. Technol.* **2019**, *30*, 2207–2220. [CrossRef]
7. Tawiah, B.; Yu, B.; Wei, R.C.; Yuen, R.K.K.; Chen, W.; Xin, J.H.; Fei, B. Simultaneous fire safety enhancement and mechanical reinforcement of poly (lactic acid) biocomposites with hexaphenyl (nitrilotris(ethane-2,1-diyl))tris (phosphoramidate). *J. Hazard. Mater.* **2019**, *380*, 120856. [CrossRef]
8. Liu, J.J.; Zhou, K.Q.; Wen, P.Y.; Wang, B.B.; Hu, Y.; Gui, Z. The influence of multiple modified MMT on the thermal and fire behavior of poly(lactic acid) nanocomposites. *Polym. Adv. Technol.* **2015**, *26*, 626–634. [CrossRef]
9. Wen, X.; Liu, Z.Q.; Li, Z.; Zhang, J.; Wang, D.Y.; Szymanska, K.; Chen, X.C.; Mijowska, E.; Tang, T. Constructing multifunctional nanofiller with reactive interface in PLA/CB-g-DOPO composites for simultaneously improving flame retardancy, electrical conductivity and mechanical properties. *Compos. Sci. Technol.* **2020**, *188*, 107988. [CrossRef]
10. Jiang, P.; Zhang, S.; Bourbigot, S.; Chen, Z.L.; Duquesne, S.; Casetta, M. Surface grafting of sepiolite with a phosphaphenanthrene derivative and its flame-retardant mechanism on PLA nanocomposites. *Polym. Degrad. Stab.* **2019**, *165*, 68–79. [CrossRef]
11. Hu, Y.D.; Xu, P.; Gui, H.G.; Wang, X.X.; Ding, Y.S. Effect of imidazolium phosphate and multiwalled carbon nanotubes on thermal stability and flame retardancy of polylactide. *Compos. Part A Appl. Sci. Manuf.* **2015**, *77*, 147–153. [CrossRef]
12. Gupta, A.; Sakthivel, T.; Seal, S. Recent development in 2D materials beyond graphene. *Prog. Mater. Sci.* **2015**, *73*, 44–126.
13. Yue, X.P.; Li, C.F.; Ni, Y.H.; Xu, Y.J.; Wang, J. Flame retardant nanocomposites based on 2D layered nanomaterials: A review. *J. Mater. Sci.* **2019**, *54*, 13070–13105. [CrossRef]
14. Tawiah, B.; Yu, B.; Fei, B. Advances in flame retardant poly(lactic acid). *Polymers* **2018**, *10*, 876. [CrossRef]
15. Cao, X.W.; Chi, X.N.; Deng, X.Q.; Liu, T.; Yu, B.; Wang, B.; Yuen, A.C.Y.; Wu, W.; Li, R.K.Y. Synergistic effect of flame retardants and graphitic carbon nitride on flame retardancy of polylactide composites. *Polym. Adv. Technol.* **2020**, *31*, 1661–1670. [CrossRef]
16. Xu, L.F.; Tan, X.W.; Xu, R.J.; Xie, J.Y.; Lei, C.H. Influence of functionalized molybdenum disulfide (MoS2) with triazine derivatives on the thermal stability and flame retardancy of intumescent Poly(lactic acid) system. *Polym. Compos.* **2019**, *40*, 2244–2257. [CrossRef]
17. Jing, J.; Zhang, Y.; Fang, Z.P.; Wang, D.Y. Core-shell flame retardant/graphene oxide hybrid: A self-assembly strategy towards reducing fire hazard and improving toughness of polylactic acid. *Compos. Sci. Technol.* **2018**, *165*, 161–167. [CrossRef]
18. Zhang, S.; Yan, Y.X.; Wang, W.J.; Gu, X.Y.; Li, H.F.; Li, J.H.; Sun, J. Intercalation of phosphotungstic acid into layered double hydroxides by reconstruction method and its application in intumescent flame retardant poly (lactic acid) composites. *Polym. Degrad. Stab.* **2018**, *147*, 142–150. [CrossRef]

19. Malvoisin, B.; Chopin, C.; Baronnet, A.; Brunet, F.; Bezacier, L.; Guillot, S. Fe-Ni-rich silicate aggregates formed after sulfides i high-pressure serpentinites. *J. Petrol.* **2017**, *58*, 963–978. [CrossRef]
20. Zhang, N.; Yang, B.P.; He, Y.Q.; He, Y.L.; Liu, X.H.; Liu, M.; Song, G.Y.; Chen, G.; Pan, A.Q.; Liang, S.Q.; et al. Serpentin $Ni_3Ge_2O_5(OH)_4$ nanosheets with tailored layers and size for efficient oxygen evolution reactions. *Small* **2018**, *14*, 180301 [CrossRef]
21. Dong, X.Y.; Zhang, Y.F.; Wang, Q.S.; Zhang, X.R.; Gao, M.; Meng, C.G. Synthesis of urchin-like $Ni_3Si_2O_5(OH)_4$ hierarchical hollov spheres/GO composite with enhanced electrochemical properties for high-performance hybrid supercapacitors. *Dalton Tran* **2019**, *48*, 11749–11762. [CrossRef] [PubMed]
22. Yan, Z.C.; Pan, T.S.; Wang, D.K.; Li, J.C.; Jin, L.; Huang, L.; Jiang, J.H.; Qi, Z.H.; Zhang, H.L.; Gao, M.; et al. Stretchabl micromotion sensor with enhanced sensitivity using serpentine layout. *ACS Appl. Mater. Inter.* **2019**, *11*, 12261–12271. [CrossRef [PubMed]
23. Liao, C.G.; Xiao, Z.Y.; Zhang, N.; Liang, B.; Chen, G.; Wu, W.; Pan, J.L.; Liu, M.; Zheng, X.R.; Kang, Q. Photo-irradiation tune highly active sites over β-$Ni(OH)_2$ nanosheets for the electrocatalytic oxygen evolution reaction. *Chem. Commun.* **2021**, *5* 9060–9063. [CrossRef] [PubMed]
24. Chen, X.F.; Huang, Y.; Han, X.P.; Zhang, K.C. Synthesis of cobalt nanofibers@nickel sulfide nanosheets hierarchical core-shel composites for anode materials of lithium ion batteries. *Electrochim. Acta* **2018**, *284*, 418–426. [CrossRef]
25. Krasilin, A.A.; Nevedomsky, V.N.; Gusarov, V.V. Comparative energy modeling of multiwalled $Mg_3Si_2O_5(OH)_4$ anc $Ni_3Si_2O_5(OH)_4$ nanoscroll growth. *J. Phys. Chem. C* **2017**, *121*, 12495–12502. [CrossRef]
26. Zheng, Y.; Zhang, P.; Yue, H.R.; Xiang, G.; Qian, Z.X.; Li, H.T.; Jiang, W.; Liang, B.; Pehkonen, S.O.; Yuan, S.J. Poly(methacryli acid)-graft-$Ni_3Si_2O_5(OH)_4$ multiwalled nanotubes as a novel nanosorbent for effective removal of copper(II) ions. *Colloids Surf.* **2016**, *502*, 89–101. [CrossRef]
27. Lu, J.W.; Sun, M.J.; Yuan, Z.T.; Qi, S.L.; Tong, Z.Y.; Li, L.X.; Meng, Q.Y. Innovative insight for sodium hexametaphosphat interaction with serpentine. *Colloids Surf. A* **2019**, *560*, 35–41. [CrossRef]
28. Xiong, Z.Q.; Zhang, Y.; Du, X.Y.; Song, P.A.; Fang, Z.P. Green and scalable fabrication of core-shell biobased flame retardants fo reducing flammability of polylactic acid. *ACS Sustain. Chem. Eng.* **2019**, *7*, 8954–8963. [CrossRef]
29. Cao, X.W.; Chi, X.N.; Deng, X.Q.; Sun, Q.J.; Gong, X.J.; Yu, B.; Yuen, A.C.Y.; Wu, W.; Li, R.K.Y. Facile synthesis of phosphoru and cobalt co-doped graphitic carbon nitride for fire and smoke suppressions of polylactide composite. *Polymers* **2020**, *12*, 1106 [CrossRef]
30. Tong, Y.Z.; Wu, W.; Zhao, W.J.; Xing, Y.R.; Zhang, H.T.; Wang, C.; Chen, T.B.Y.; Yuen, A.C.Y.; Yu, B.; Cao, X.W.; et al. Nanohybri of Co_3O_4 nanoparticles and polyphosphazene-decorated ultra-thin boron nitride nanosheets for simultaneous enhancement ir fire safety and smoke suppression of thermoplastic polyurethane. *Polymers* **2022**, *14*, 4341. [CrossRef]
31. Wu, W.; Huang, W.J.; Tong, Y.Z.; Huang, J.S.; Wu, J.C.; Cao, X.W.; Zhang, Q.C.; Yu, B.; Li, R.K.Y. Self-assembled double core shell structured zeolitic imidazole framework-8 as an effective flame retardant and smoke suppression agent for thermoplastic polyurethane. *Appl. Surf. Sci.* **2022**, *610*, 155540. [CrossRef]
32. Yu, H.L.; Cui, J.F.; Zhang, H.Y.; Yang, B.P.; Guo, J.H.; Mu, B.; Wang, Z.H.; Li, H.M.; Tian, L.L. A novel flame retardant consisting of functionalized Salen-Ni based polyphosphazene microspheres. *High Perform. Polym.* **2022**, *34*, 09540083221094972. [CrossRef]
33. Wang, X.G.; Wang, S.H.; Wang, W.J.; Li, H.F.; Liu, X.D.; Gu, X.Y.; Bourbigot, S.; Wang, Z.W.; Sun, J.; Zhang, S. The flammability and mechanical properties of poly (lactic acid) composites containing Ni-MOF nanosheets with polyhydroxy groups. *Compos Part B Eng.* **2020**, *183*, 107568. [CrossRef]
34. Wu, Q.; Cui, X.Y.; Mu, C.Z.; Sun, J.; Gu, X.Y.; Li, H.F.; Zhang, S. Toward a new approach to synchronously improve the fire performance and toughness of polylactic acid by the incorporation of facilely synthesized ammonium polyphosphate derivatives *Compos. Part A Appl. Sci.* **2021**, *150*, 106595.
35. Li, D.F.; Zhao, X.; Jia, Y.W.; He, L.; Wang, X.L.; Wang, Y.Z. Dual effect of dynamic vulcanization of biobased unsaturated polyester Simultaneously enhance the toughness and fire safety of Poly(lactic acid). *Compos. Part B Eng.* **2019**, *175*, 107069. [CrossRef]
36. Cai, W.; Cai, T.; He, L.; Chu, F.; Mu, X.; Han, L.; Hu, Y.; Wang, B.; Hu, W. Natural antioxidant functionalization for fabricating ambient-stable black phosphorus nanosheets toward enhancing flame retardancy and toxic gases suppression of polyurethane *J. Hazard Mater.* **2020**, *387*, 121971.
37. Han, D.Q.; Wang, H.; Lu, T.T.; Cao, L.Y.; Dai, Y.F.; Cao, H.Z.; Yu, X.L. Scalable manufacturing green core–shell structure flame retardant, with enhanced mechanical and flame-retardant performances of polylactic acid. *J. Polym. Environ.* **2022**, *30*, 2516–2533. [CrossRef]
38. Zhao, H.; Gao, W.C.; Li, Q.; Khan, M.R.; Hu, G.H.; Liu, Y.; Wu, W.; Huang, C.X.; Li, R.K.Y. Recent advances in superhydrophobic polyurethane: Preparations and applications. *Adv. Colloid Interface Sci.* **2022**, *303*, 102644. [CrossRef]
39. Ye, G.F.; Huo, S.Q.; Wang, C.; Shi, Q.; Liu, Z.T.; Wang, H. One-step and green synthesis of a bio-based high-efficiency flame retardant for poly (lactic acid). *Polym. Degrad. Stab.* **2021**, *192*, 109696. [CrossRef]
40. Wu, W.; Zhao, W.; Gong, X.; Sun, Q.; Cao, X.; Su, Y.; Yu, B.; Li, R.K.; Vellaisamy, R.A. Surface decoration of halloysite nanotubes with POSS for fire-safe thermoplastic polyurethane nanocomposites. *J. Mater. Sci. Technol.* **2022**, *101*, 107–117. [CrossRef]
41. Zhan, Y.Y.; Wu, X.J.; Wang, S.S.; Yuan, B.H.; Fang, Q.; Shang, S.; Cao, C.R.; Chen, G.Q. Synthesis of a bio-based flame retardant via a facile strategy and its synergistic effect with ammonium polyphosphate on the flame retardancy of polylactic acid. *Polym. Degrad. Stab.* **2021**, *191*, 109684. [CrossRef]

42. Wang, D.; Wang, Y.Z.; Zhang, X.H.; Li, T.; Du, M.L.; Chen, M.Q.; Dong, W.F. Preferred zinc-modified melamine phytate for the flame retardant polylactide with limited smoke release. *New J. Chem.* **2021**, *45*, 13329–13339. [CrossRef]
43. Jin, X.D.; Cui, S.P.; Sun, S.B.; Gu, X.Y.; Li, H.F.; Liu, X.D.; Tang, W.F.; Sun, J.; Bourbigot, S.; Zhang, S. The preparation of a bio-polyelectrolytes based core-shell structure and its application in flame retardant polylactic acid composites. *Compos. Part A Appl. Sci.* **2019**, *124*, 105485. [CrossRef]
44. Wu, W.; Wu, C.K.; Peng, H.Y.; Sun, Q.J.; Zhou, L.; Zhuang, J.Q.; Cao, X.W.; Roy, V.A.L.; Li, R.K.Y. Effect of nitrogen-doped graphene on morphology and properties of immiscible poly(butylene succinate)/polylactide blends. *Compos. Part B Eng.* **2017**, *113*, 300–307. [CrossRef]
45. Zhao, H.B.; Cui, Z.X.; Wang, X.F.; Turng, L.S.; Peng, X.F. Processing and characterization of solid and microcellular poly(lactic acid)/polyhydroxybutyrate-valerate (PLA/PHBV) blends and PLA/PHBV/Clay nanocomposites. *Compos. Part B Eng.* **2013**, *51*, 79–91. [CrossRef]
46. Zhao, H.; Li, K.C.; Wu, W.; Li, Q.; Jiang, Y.n.; Cheng, B.X.; Huang, C.X.; Li, H.N. Microstructure and viscoelastic behavior of waterborne polyurethane/cellulose nanofiber nanocomposite. *J. Ind. Eng. Chem.* **2022**, *110*, 150–157. [CrossRef]

polymers

MDPI

Article

Tuning the Ferroelectric Response of Sandwich-Structured Nanocomposites with the Coordination of $Ba_{0.6}Sr_{0.4}TiO_3$ Nanoparticles and Boron Nitride Nanosheets to Achieve Excellent Discharge Energy Density and Efficiency

Zhihui Yi, Zhuo Wang *, Dan Wu and Ying Xue

Shaanxi Key Laboratory of Green Preparation and Functionalization for Inorganic Materials, School of Materials Science and Engineering, Shaanxi University of Science & Technology, Xi'an 710021, China

* Correspondence: wangzhuo@sust.edu.cn

Abstract: With the rapid development of new electronic products and sustainable energy systems, there is an increasing demand for electrical energy storage devices such as electrostatic capacitors. In order to comprehensively improve the dielectric, insulating, and energy storage properties of PVDF-based composites, sandwich-structured composites were prepared by layer-by-layer solution casting. The outer layers of the sandwich structure composite are both PVDF/boron nitride nanosheet composites, and the middle layer is a PVDF/$Ba_{0.6}Sr_{0.4}TiO_3$ nanoparticles composite. The structural and electrical properties of the sandwich-structured composites were characterized and analyzed. The results show that when the volume percentage of $Ba_{0.6}Sr_{0.4}TiO_3$ nanoparticles in the middle layer of the sandwich structure composite is 1 vol.%, the dielectric properties are significantly improved. Its dielectric constant is 8.99 at 10 kHz, the dielectric loss factor is 0.025, and it has better insulating properties and resistance to electrical breakdown. Benefiting from the high breakdown electric field strength and the large maximum electrical displacement, the sandwich-structured composites with 1 vol.% and $Ba_{0.6}Sr_{0.4}TiO_3$ nanoparticles in the middle layer show a superior discharge energy density of 8.9 J/cm^3, and excellent charge and discharge energy efficiency of 76%. The sandwich structure composite achieves the goal of simultaneous improvement in breakdown electric field strength and dielectric constant.

Keywords: PVDF; nanocomposites; energy storage; $Ba_{0.6}Sr_{0.4}TiO_3$ NPs; dielectrics

Citation: Yi, Z.; Wang, Z.; Wu, D.; Xue, Y. Tuning the Ferroelectric Response of Sandwich-Structured Nanocomposites with the Coordination of $Ba_{0.6}Sr_{0.4}TiO_3$ Nanoparticles and Boron Nitride Nanosheets to Achieve Excellent Discharge Energy Density and Efficiency. *Polymers* **2023**, *15*, 3642. https://doi.org/10.3390/polym15173642

Academic Editors: Hui Zhao, Wei Wu and Bin Wang

Received: 30 June 2023
Revised: 22 August 2023
Accepted: 24 August 2023
Published: 4 September 2023

1. Introduction

Driven by the rapidly depleted background of limited fossil fuels, a great deal of research has been focused on the exploration of renewable, green, and sustainable energy sources and advanced energy storage technologies [1,2]. Compared with batteries, fuel cells, and supercapacitors, dielectric capacitors stand out among various energy storage devices because of their ultra-fast charge-discharge speed, high power density, and longer cycle times [3]. These key properties have greatly facilitated the widespread application of dielectric capacitors in modern electronic and electrical industries, including but not limited to portable electronic devices, computing systems, high-power pulsed lasers, and smart grids.

The charge and discharge energy density of a dielectric capacitor are determined by the dielectric material. In order to improve the energy storage density of capacitors, the method of compounding organic/inorganic materials can be used to prepare polymer-based composite dielectric materials [4]. As a new type of dielectric material, polymer-based composite dielectric materials can combine the unique thermal, mechanical, and electrical properties of inorganic materials with the good processability and high breakdown strength of polymer materials [5]. According to the calculation Formula (1) of the energy storage density of dielectric material [6]:

$$U = \int_0^{D_{max}} EdD = \int_0^{P_{max}} EdP \quad (1)$$

The energy storage density value of dielectric material is determined by the electric field strength and electric displacement in the charge-discharge curve [7]. Therefore, in order to obtain materials with high energy storage density, on the one hand, it is necessary to increase the breakdown electric field strength of the dielectric material as much as possible [8]. In order to improve the dielectric constant of dielectric materials, high dielectric ceramic fillers are often introduced into polymers in large quantities [9–11]. Unfortunately, due to the huge difference between the dielectric constant of the high dielectric filler and the polymer matrix, the local electric field distribution of the composite material is uneven, and the breakdown strength of the composite material is greatly reduced [12,13].

In order to improve the high electric field resistance of composite materials, a large number of inorganic two-dimensional materials such as montmorillonite nanosheets, zirconia nanosheets, sodium bismuth titanate nanosheets, potassium sodium niobate nanosheets, and kaolin nanosheets are used as inorganic fillers in the polymer matrix [14–16]. The insulating two-dimensional inorganic material forms a curved channel when the electrical dendrites inside the composite material evolve in the polymer matrix, which increases the tortuosity of the breakdown path, thereby improving the breakdown field strength of the composite material [17,18]. The two-dimensional sheet-like inorganic materials can generate more traps, which can effectively scatter the charges injected into the composite material [19]. The two-dimensional sheet-like inorganic material reduces the leakage current value of the material, thereby reducing the conductivity loss, which is beneficial to the improvement in the energy storage density and the charge-discharge efficiency of the dielectric material [20]. Although the addition of these 2D inorganic materials improves the breakdown field strength of composite dielectrics, the dielectric constants of the resulting composites are usually not high, and some are even lower than pure ferroelectric fluoropolymers [21]. The low dielectric constant limits the improvement in the energy storage density of composite materials to a greater extent [22].

To solve the above problems, composite materials with a sandwich structure have been designed [23–25]. Compared with single-layer composite films, composite films with a sandwich structure are expected to solve the contradiction that composite materials composed of inorganic and organic materials cannot simultaneously achieve improved breakdown strength and improved high dielectric constant [26,27]. Layers of materials with high dielectric constants and high breakdown strength are stacked layer by layer to form sandwich-structured composites [28,29]. In sandwich-structured composites, different functional layers exhibit synergistic advantages in enhancing energy density [30,31].

In this paper, polyvinylidene fluoride (PVDF) was chosen as the polymer matrix of the sandwich-structured composites. At a frequency of 100 Hz, the dielectric constant of PVDF is close to 10, which is beneficial to the improvement in the dielectric constant and dielectric polarization strength of the composite material [32]. $Ba_{0.6}Sr_{0.4}TiO_3$ nanoparticles ($Ba_{0.6}Sr_{0.4}TiO_3$ NPs, BST NPs) with a high dielectric constant [33], and two-dimensional boron nitride nanosheets (BNNS) are used as inorganic fillers in the middle and outer layers, respectively. Compared with other inorganic particles, barium strontium titanate nanoparticles have a higher dielectric constant, which contributes more to the improvement in the dielectric constants of composite materials. BNNS has high forbidden bandwidth (6 eV), high insulation, high breakdown electric field strength (800 kV/mm), and high thermal conductivity [34–36]. The addition of BNNS is expected to improve the insulating properties of the materials and thus improve the dielectric energy storage properties of the composites.

2. Experimental Section

2.1. Raw Materials

Boron nitride nanosheets (BNNS), acetylacetone, glacial acetic acid, N, N-dimethylformamide (DMF), anhydrous ethanol, boron nitride nanosheets, dopamine hydrochloride (DA-HCl), oxalic acid dihydrate ($HC_2O_4 \cdot 2H_2O$), barium nitrate ($Ba(NO_3)_2$), strontium nitrate ($Sr(NO_3)_2$), and tetrabutyl titanate were purchased from Sinopharm Chemical Reagent Co., Ltd., Shanghai, China. Polyvinylidene fluoride (PVDF) was obtained from Sigma-Aldrich Co., LLC, St. Louis, MO, USA.

2.2. Fabrication of $Ba_{0.6}Sr_{0.4}TiO_3$ Nanoparticles

To prepare $Ba_{0.6}Sr_{0.4}TiO_3$ nanoparticles, $Ba(NO_3)_2$, tetrabutyl titanate, and $Sr(NO_3)_2$ were used as the main raw materials, oxalic acid dihydrate was used as precipitant, and a mixed solution (deionized water and ethanol) was used as solvent. The specific experimental steps were as follows: A beaker was poured with 0.1 mol of $HC_2O_4 \cdot 2H_2O$. Amounts of 150 mL of distilled water and 50 mL of absolute ethanol were measured as solvents. Under the condition of heating in a water bath, the liquid in the beaker was stirred, so that $HC_2O_4 \cdot 2H_2O$ was completely dissolved, and the mixed solution was marked as Mixture A. An amount of 0.05 mol of tetrabutyl titanate was dissolved in 100 mL of absolute ethanol, heated and stirred to mix well, and this mixed solution was marked as Mixture B. Amounts of 0.02 mol $Sr(NO_3)_2$ and 0.03 mol $Ba(NO_3)_2$ were dissolved in 100 mL of distilled water, heated and stirred until $Sr(NO_3)_2$ and 0.03 mol $Ba(NO_3)_2$ were completely dissolved, and the mixed solution was marked as a Mixture C. Mixture B and Mixture A were mixed, and after being stirred evenly, ammonia water was added to adjust the pH to about 3.5 to obtain Mixture D. Under the condition of heating to 80 °C in a water bath, Mixture C was slowly added dropwise to Mixture D, and then stirred for 2 h to separate the compounds. After the complete reaction, the resulting mixed solution was aged for 24 h. After being aged for 24 h, the precursor of $Ba_{0.6}Sr_{0.4}TiO_3$ powder was centrifuged and washed. Then, the precursor of $Ba_{0.6}Sr_{0.4}TiO_3$ powder was dried in a drying oven at 100 °C for 6 h and calcined in a muffle furnace at 800 °C. After the $Ba_{0.6}Sr_{0.4}TiO_3$ powder was ground by a ball mill for 12 h, the $Ba_{0.6}Sr_{0.4}TiO_3$ nanopowders were successfully prepared. A schematic diagram of the fabrication of $Ba_{0.6}Sr_{0.4}TiO_3$ nanoparticles is shown in Figure S1.

2.3. Preparation of $Ba_{0.6}Sr_{0.4}TiO_3$@DA Nanoparticles

First, the surface of BST was treated with hydroxyl groups. The $Ba_{0.6}Sr_{0.4}TiO_3$ nanopowders were dispersed in a mixed solution (ethanol and water) with a ratio of ethanol to water of 1:1, stirred and ultrasonically dispersed for 4 h, and dried for 12 h. Then, the treated $Ba_{0.6}Sr_{0.4}TiO_3$ nanopowders were added to a dopamine buffer solution (dopamine hydrochloride), and Mixture E was obtained by ultrasonic stirring for 10 h. Finally, Mixture E was centrifuged and dried, and the $Ba_{0.6}Sr_{0.4}TiO_3$@DA nanoparticles were successfully obtained. Chemical surface modification of inorganic nanoparticles can improve the uniformity of nanocomposites [37]. The distribution of the diameters of $Ba_{0.6}Sr_{0.4}TiO_3$ nanoparticles is shown in Figure S2. The diameter of the filler is mainly distributed in the 800 nm attachment. The TEM images of $Ba_{0.6}Sr_{0.4}TiO_3$@DA NPs are shown in Figure S3.

2.4. Preparation of Boron Nitride Nanosheets

Boron nitride nanosheets (BNNS) were obtained by exfoliating hexagonal boron nitride powder (h-BN) by the chemical solvent ultrasonic method. The preparation process is shown in Figure 1. An amount of 1 g of hexagonal boron nitride powder was accurately weighed into a centrifuge tube, and absolute ethanol was added to the centrifuge tube. The centrifuge tube was placed under the ultrasonic probe, the height of the ultrasonic probe was adjusted, and the probe made as deep as possible toward the bottom of the centrifuge tube, without touching the wall of the centrifuge tube. The centrifuge tube mouth was sealed with parafilm to prevent solvent volatilization during ultrasonication. The power of the sonicator was set to 200 W. The mixture in the centrifuge tube was centrifuged after 2 h of uninterrupted operation of the needle tip sonicator. After centrifugation, the supernatant was discarded and the bottom pellet collected. The obtained precipitate was placed in a vacuum drying oven at 60 °C for 24 h to obtain boron nitride nanosheets (BNNS).

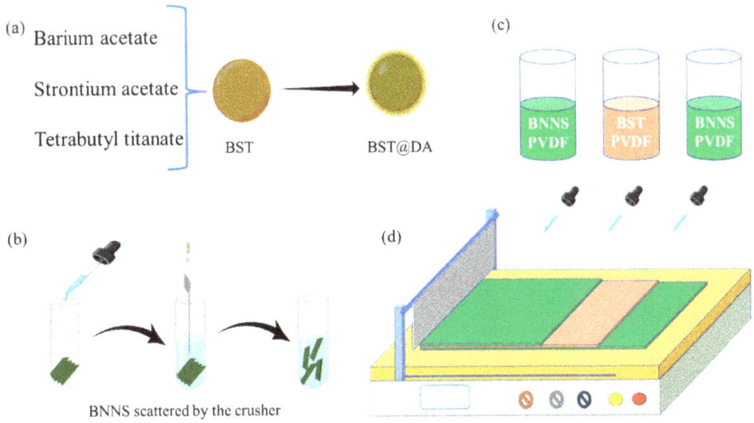

Figure 1. Schematic diagram of the preparation process of nanocomposite films. (**a**) Preparation of nano-filler, (**b**) Preparation of boron nitride nanosheets, (**c**) Preparation of mixed solution for tape casting, (**d**) The steps of casting method.

2.5. Preparation of Sandwich-Structured Composite Films

The sandwich-structured composite film material was prepared by way of layer-by-layer solution casting. The fabrication process of the composite films is shown in Figure 1. In this paper, 5 wt.% PVDF/BNNS-0 vol.% PVDF/$Ba_{0.6}Sr_{0.4}TiO_3$@DA-5 wt.% PVDF/BNNS can be abbreviated as B0B, 5 wt.% PVDF/BNNS-1 vol.% PVDF/$Ba_{0.6}Sr_{0.4}TiO_3$@DA-5 wt.% PVDF/BNNS can be abbreviated as B1B, 5 wt.% PVDF/BNNS-2 vol.% PVDF/$Ba_{0.6}Sr_{0.4}TiO_3$ @DA-5 wt.% PVDF/BNNS can be abbreviated as B2B, and 5 wt.% PVDF/BNNS-3 vol.% PVDF/$Ba_{0.6}Sr_{0.4}TiO_3$@DA-5 wt.% PVDF/BNNS can be abbreviated as B3B.

2.6. Characterization

In this study, the phase structures of inorganic nanoparticles and nanocomposites were analyzed using an X-ray diffractometer (XRD). The model of this X-ray diffractometer is Rigaku D/max-2200. The microstructure of the samples was analyzed by field emission scanning electron microscopy (FEI). The composition of the samples was analyzed using an energy dispersive spectrometer (EDS). The specific sample preparation method is as follows: For powder SEM sample preparation, the powder was dissolved in ethanol, and ultrasonic shock was performed for 30 min. The upper liquid was taken from the capillaries on the aluminum foil, dried and sprayed with gold, and observed. For sandwich-structured composite material SEM sample preparation, the composite material was placed in liquid nitrogen for quenching, the quenched sample was placed on the sample table, and the microstructure was observed after drying and spraying gold. Gold in a circle shape with a diameter of 1 cm was sputtered onto both sides of the composite films as the electrodes for the dielectric characterization. The dielectric constant and loss of the composite films were characterized using an Agilent E4980A. An Agilent precision impedance analyzer (E4980A, Agilent, Santa Clara, CA, USA) was also used for AC conductivity tests. The test frequency was 20 Hz-2 MHz, the temperature was room temperature, and the measurement circuit adopted a four terminal double lead mode. The AC conductivity of the sample was obtained by using the instrument to test the impedance of the sample. Gold in a circle shape with a diameter of 2 mm was sputtered onto both sides of the composite films as the electrodes for the characterization of ferroelectric parametron properties. The D-E hysteresis loops of the composite films were determined using a Radiant Premier II. A breakdown strength test was performed using a Precision Multiferroic ferroelectric tester from Radiant Technologies, USA. The test temperature range was room temperature, and the test frequency was 10 Hz.

3. Results and Discussion

It can be seen from Figure 2a that BNNS mainly has a strong diffraction peak at 27°, which corresponds to the diffraction position of the (002) crystal plane. The peak is strong and sharp, indicating that the crystals of boron nitride nanosheets obtained after exfoliation are still highly ordered. The positions of the XRD peaks of the BST nanoparticles in Figure 2a are in perfect agreement with the standard $Ba_{0.6}Sr_{0.4}TiO_3$ card, which proves that the perovskite-structured barium strontium titanate was successfully synthesized. No other peaks were generated in the XRD spectrum, indicating that the prepared BST nanoparticles were of high purity.

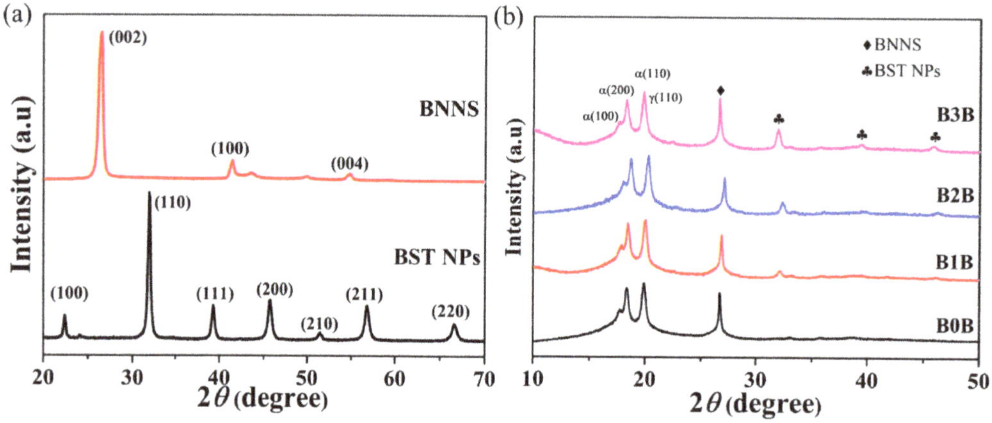

Figure 2. (**a**) XRD patterns of as-prepared $Ba_{0.6}Sr_{0.4}TiO_3$ NPs and as-prepared BNNS. (**b**) XRD patterns of as-prepared different nanocomposites.

In order to analyze the phase structure of the sandwich-structured nanocomposites, the sandwich-structured nanocomposites with contents of 0 vol.%, 1 vol.%, 2 vol.%, and 3 vol.% of BST nanoparticles in the interlayer were characterized by X-ray diffraction, as shown in Figure 2b. The nano-nanocomposite material also appeared with a new diffraction peak, that is, the characteristic diffraction peak of BNNS at 27°, indicating that the crystalline structure of PVDF and BNNS was not destroyed in the composite of PVDF and BNNS. As can be seen from Figure 2b, for the prepared three-layer composite material, in addition to the diffraction peaks of PVDF, the characteristic diffraction peaks belonging to BST and the characteristic diffraction peaks of BNNS appear at the same time. It was demonstrated that sandwich-structured nanocomposite films were successfully prepared without significant damage to the crystallinity of BNNS and BST. The crystalline phase of PVDF in sandwich composites is mainly composed of the non-polar α phase and γ phase [38], in which the γ phase mainly originates from the quenching process of the composite film instantaneously transferred from a high temperature of 200 °C to ice water. The existence of nonferroelectric α and γ phases is beneficial to the dipole rotation of the composite film during the removal of the electric field, which helps to reduce the remnant polarization.

The SEM results of $Ba_{0.6}Sr_{0.4}TiO_3$ NPs are shown in Figure 3a. The average diameter of $Ba_{0.6}Sr_{0.4}TiO_3$ NPs is 500 nm, which means that the $Ba_{0.6}Sr_{0.4}TiO_3$ NPs used in this paper can provide larger interfacial polarization and a higher dielectric constant. At the same time, they can also meet the high dielectric constant filler requirements for constructing high dielectric constant layers in "sandwich" composite dielectrics. The microscopic morphology of BNNS can be seen from the SEM image in Figure 3b. BNNS exhibits a lamellar structure, and the size in the plane direction is about 100 nm. The good dispersibility of BNNS is attributed to the strong polarity caused by the B-N chemical bond on the molecular plane, which not only helps its uniform dispersion in DMF but also its uniform distribution in the polar polymer matrix.

Figure 3. The SEM of (**a**) $Ba_{0.6}Sr_{0.4}TiO_3$ NPs and (**b**) BNNS.

A schematic diagram of the sandwich-structured PVDF-based nanocomposite dielectric material is shown in Figure 4. Figure 4a,d,g,j are a cross-sectional scan of the sandwich-structured nanocomposite. The contents of BST NPs in the middle layer are 0 vol.%, 1 vol.%, 2 vol.%, and 3 vol.%, respectively. It can be seen from the cross-sectional scan that the nanofillers have good dispersibility in the polymer matrix of each layer. There are no defects such as separation or voids at the interface between different layers, indicating that the preparation process is good. There is no introduction of defects such as air, which is beneficial to the improvement in the dielectric properties of the three-layer structure material. The layered structure could not be clearly observed from the quenched cross-section of the sandwich composite film, so we performed an EDS surface scanning test on the sandwich composite. The distribution of the main elements (N, Ti) in the material is clearly observed, and the results are shown in Figure 4b–f,h,i,k,l, showing a uniform layered distribution. The Ti element is concentrated in the middle part of the sandwich composite film. The N element (BNNS) is dispersed on both sides of the sandwich composite film. With the increasing volume percentage of $Ba_{0.6}Sr_{0.4}TiO_3$ NPs in the interlayer, the Ti element is more and more enriched, which is completely consistent with the original design of the experiment.

The relationship between the dielectric constant and the dielectric loss as a function of frequency is characterized by the sandwich structure composite when the content of $Ba_{0.6}Sr_{0.4}TiO_3$ NPs in the intermediate layer changes, as shown in Figure 5. As the content of $Ba_{0.6}Sr_{0.4}TiO_3$ NPs in the interlayer gradually increased, the dielectric constant of the sandwich structure material also gradually increased. At 1000 Hz, the dielectric constant of $Ba_{0.6}Sr_{0.4}TiO_3$ NPs with 3 vol.% volume fraction added is 10.71. The increase in the dielectric constant is mainly due to the high dielectric constant of $Ba_{0.6}Sr_{0.4}TiO_3$ NPs. It can be seen from Figure 5b that although the content of $Ba_{0.6}Sr_{0.4}TiO_3$ NPs in the interlayer is added to 3 vol.%, the dielectric loss of the sandwich composites is always maintained at a low level ($\tan\delta < 0.04$).

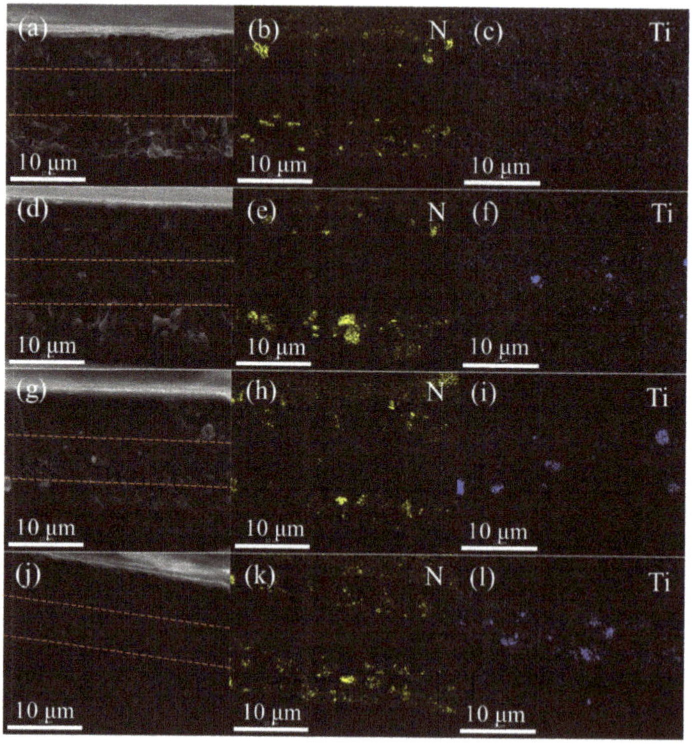

Figure 4. (**a**,**d**,**g**,**j**) Cross-sectional SEM image of the sandwich-structured nanocomposite with 0 vol.%, 1 vol.%, 2 vol.%, and 3 vol.% $Ba_{0.6}Sr_{0.4}TiO_3$ NPs in the central layer, respectively. (**b**,**e**,**h**,**k**) EDS element (N) distribution of the sandwich-structured nanocomposite with 0 vol.%, 1 vol.%, 2 vol.%, and 3 vol.% $Ba_{0.6}Sr_{0.4}TiO_3$ NPs in the central layer, respectively. (**c**,**f**,**i**,**l**) EDS element (Ti) of the sandwich-structured nanocomposite with 0 vol.%, 1 vol.%, 2 vol.%, and 3 vol.% $Ba_{0.6}Sr_{0.4}TiO_3$ NPs in the central layer, respectively.

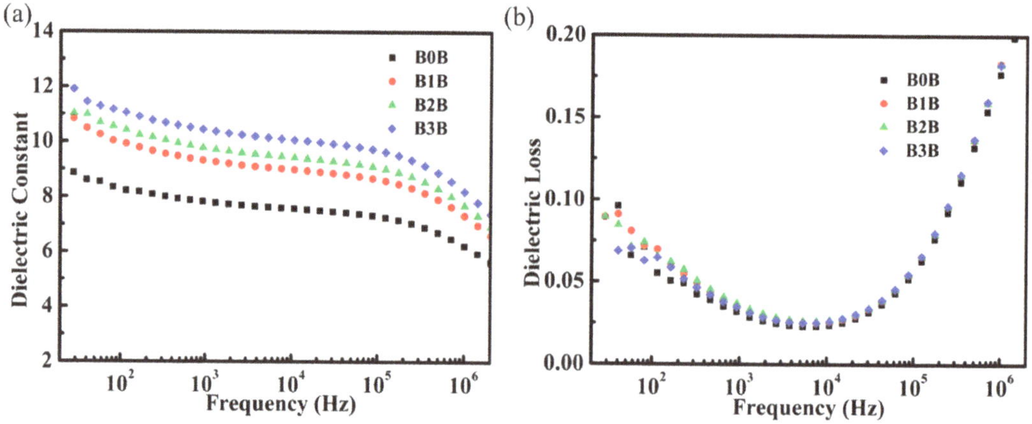

Figure 5. Frequency dependence of (**a**) dielectric constant and (**b**) dielectric loss of composites at room temperature.

Figure 6 shows the AC conductivity of sandwich-structured composite films. It can be seen from Figure 6a that the AC conductivities of the sandwich-structured composites are very low because BNNS can hinder the injection of charges at the electrodes and hinder the transport of charges in the polymer matrix. Low AC conductivity can reduce the leakage current of sandwich composites under a high field. On the other hand, the smaller conductivity loss of the sandwich composite will reduce the remanent polarization and increase the breakdown field strength of the composite. As shown in Figure 6b, the AC conductivity of the sandwich-structured composites exhibits a strong frequency dependence, which increases nonlinearly with increasing frequency. The conductivity of sandwich-structured composites increased with the increase in $Ba_{0.6}Sr_{0.4}TiO_3$ NPs loading in the middle layer. However, the increase in conductivity of the sandwich-structured composites did not change by an order of magnitude. All sandwich-structured composites still have good insulating properties.

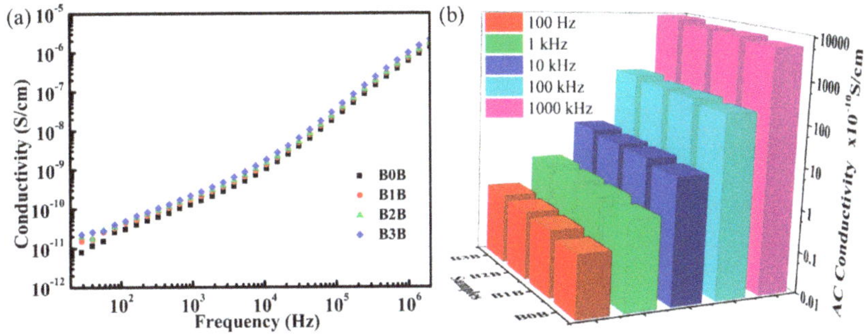

Figure 6. (**a**) AC conductivity of sandwich-structured nanocomposites with 0 vol.%, 1 vol.%, 2 vol.%, and 3 vol.% $Ba_{0.6}Sr_{0.4}TiO_3$ NPs in the central layer. (**b**) Comparison of the AC conductivities of sandwich-structured composites with different $Ba_{0.6}Sr_{0.4}TiO_3$ NP contents in the middle layer at 100 Hz, 1 kHz, 10 kHz, 100 kHz, and 1000 kHz.

The electrical displacement and electric field intensity curves were tested for pure PVDF polymer and sandwich nanocomposites. Figure 7a shows the D-E curve measured at an electric field strength of 200 kV/mm. It can be seen from Figure 7a that the maximum electrical displacement value of B0B at 200 kV/mm is only 3.72027 μC/cm^2, which is due to the low dipole polarizability caused by PVDF with a low dielectric constant. With the increase in the content of $Ba_{0.6}Sr_{0.4}TiO_3$ NPs in the interlayer, the maximum electrical displacement value gradually increased. At the addition of 3 vol.% $Ba_{0.6}Sr_{0.4}TiO_3$ NPs, the maximum electrical displacement is 5.86947 μC/cm^2, which is due to the gradual increase in the dielectric constant of the sandwich composites with the increase in $Ba_{0.6}Sr_{0.4}TiO_3$ NPs. The remanent polarization of the sandwich structure also increases with the content of $Ba_{0.6}Sr_{0.4}TiO_3$ NPs. At 0 vol.% $Ba_{0.6}Sr_{0.4}TiO_3$ NP content, the remnant polarization is 0.31132 μC/cm^2. With the gradual increase in the $Ba_{0.6}Sr_{0.4}TiO_3$ NP content in the interlayer, the remanent polarization also increases gradually. At 1 vol.% $Ba_{0.6}Sr_{0.4}TiO_3$ NP content, the remnant polarization is 0.39749 μC/cm^2. When the addition of $Ba_{0.6}Sr_{0.4}TiO_3$ NPs was 3 vol.%, the remnant polarization also reached the maximum value of 0.80512 μC/cm^2, and the overall low remnant polarization of the sandwich composites is mainly attributed to the hindering effect of the outer BNNS. The existence of the outer layer of BNNS can effectively hinder the injection of electrons, thereby reducing the leakage current value of the material, which is also reflected in the AC conductivity of the composite material.

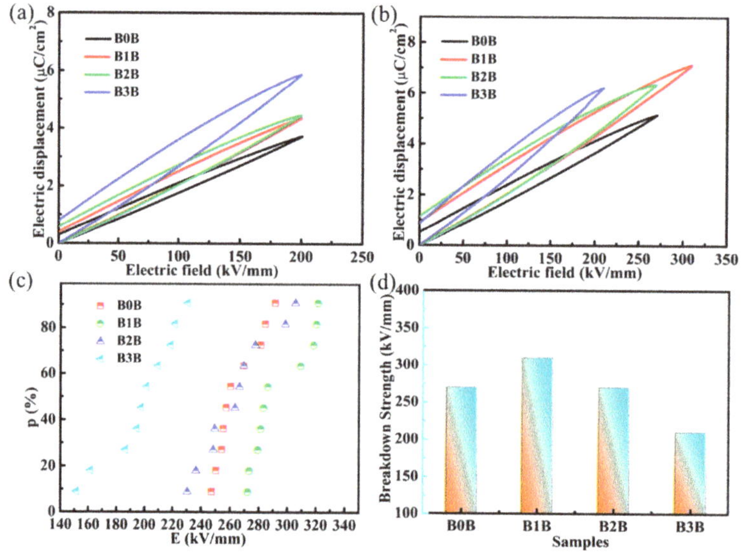

Figure 7. (**a**) Comparison of D-E loops of a series of sandwich-structured nanocomposites as a function of $Ba_{0.6}Sr_{0.4}TiO_3$ NP content at an electric field of 200 kV/mm. (**b**) Comparison of D-E loops of a series of sandwich-structured nanocomposites as a function of $Ba_{0.6}Sr_{0.4}TiO_3$ NP content at breakdown field strength. (**c**) Weibull plots of the sandwich-structured nanocomposites at varied $Ba_{0.6}Sr_{0.4}TiO_3$ NP contents in the middle layer, (**d**) Weibull breakdown strength of the sandwich-structured nanocomposites as a function of $Ba_{0.6}Sr_{0.4}TiO_3$ NP content in the middle layer.

The D-E curve of the composite under the maximum electric field is shown in Figure 7b. It can be seen from Figure 7b that the sandwich-structured composite exhibits a nonlinear (ferroelectric) response to the external electric field at high fields, which is mainly due to the inherent ferroelectricity of PVDF and $Ba_{0.6}Sr_{0.4}TiO_3$ NPs. The breakdown field strength of the prepared sandwich-structured nanocomposites also was analyzed, and the two-parameter Weibull distribution method was used for fitting [39].

$$P(E) = 1 - \exp\left[-\left(\frac{E}{E_b}\right)^{\beta}\right] \quad (2)$$

In the above Formula (2), the value of $P(E)$ is the cumulative failure probability of the electric field, and E_b and E, respectively, refer to the breakdown field strength and the test breakdown field strength when the cumulative failure probability is 63.2% [40]. β is a shape parameter related to the discreteness of the experimental data. A high β value indicates that the nanocomposites are of high quality. The Weibull distribution diagram of the breakdown field strength is shown in Figure 7c, and Figure 7d is the variation of the Weibull breakdown field strength value with the content of $Ba_{0.6}Sr_{0.4}TiO_3$ NPs in the intermediate layer. It can be seen from the figure that with the increase in $Ba_{0.6}Sr_{0.4}TiO_3$ NPs in the sandwich structure material, the breakdown field strength of the three-layer composite material increases first and then decreases. When the $Ba_{0.6}Sr_{0.4}TiO_3$ NP content was increased to 1 vol.%, the breakdown field strength of the three-layer composite was the largest, at 310 kV/mm. As the content of $Ba_{0.6}Sr_{0.4}TiO_3$ NPs continued to increase, the breakdown field strength of the composites gradually decreased. With the highest addition of $Ba_{0.6}Sr_{0.4}TiO_3$ NPs (3 vol.%), the breakdown field strength of the composite dropped to 210 kV/mm. With the increase in the content of $Ba_{0.6}Sr_{0.4}TiO_3$ NPs in the middle layer, the sandwich structure composites showed a trend of first increasing and then decreasing: One reason is that after the content of $Ba_{0.6}Sr_{0.4}TiO_3$ NPs exceeds 1 vol.%, the volume fraction of nanofillers may exceed the ideal percentage content that the polymer matrix can accommodate when the inorganic fillers are uniformly distributed. Too much filler leads to uneven distribution of $Ba_{0.6}Sr_{0.4}TiO_3$ NPs in the PVDF matrix, or even agglomeration, which affects the improvement in the breakdown field strength. Another reason is the voltage divider between the different layers, which has been similarly reported in the work of others. In sandwich-structured nanocomposites composed of different dielectric layers, the electric field strength of the low-k layers is higher than that of the high-k layers. For the sandwich-structured nanocomposites in this study, the outer layer PVDF/BNNS can withstand higher electric field strengths than the middle layer PVDF/$Ba_{0.6}Sr_{0.4}TiO_3$ NPs with a high dielectric constant due to its lower dielectric constant. When the overall three-layer composite is under the same voltage, with the increase in the $Ba_{0.6}Sr_{0.4}TiO_3$ NP content in the intermediate layer, the dielectric constant of the intermediate layer increases continuously, so the voltage shared by the intermediate layer gradually decreases. When the content of $Ba_{0.6}Sr_{0.4}TiO_3$ NPs is less than 1 vol.%, the intermediate layer will be broken down first due to the existence of $Ba_{0.6}Sr_{0.4}TiO_3$ NPs, and the withstand voltage characteristics will be reduced. When the content of $Ba_{0.6}Sr_{0.4}TiO_3$ NPs was greater than 1 vol.%, the voltage obtained by the outer layer continued to increase. Although the outer two layers have improved withstand voltage characteristics due to the presence of BNNS, the pressure drop has exceeded the withstand voltage range, causing the outer two layers to be broken down. The breakdown field strength of the final sandwich nanocomposite reaches the maximum value when the $Ba_{0.6}Sr_{0.4}TiO_3$ NP content is 1 vol.%.

In addition to the above-mentioned breakdown field strength and polarization properties, D_m-D_r also plays an important role in the performance of dielectric energy storage. Thus, the D_m, D_r, and D_m-D_r of the complex were also calculated from the D-E curve, as shown in Figure 7a. Compared to B0B, B2B, and B3B, the D_m of the sandwich-structured composites is significantly improved due to the significantly higher breakdown field strength of B1B. At the same time, the D_r of B1B is kept at a very low level. Lower D_r is beneficial for high discharge energy density because it means that the stored energy can be effectively released during the discharge process. With the increase in $Ba_{0.6}Sr_{0.4}TiO_3$ NP content, the D_m-D_r of sandwich-structured composites first increased and then decreased. It should be noted that B1B exhibited the highest D_m-D_r (6.2 C/cm^2). The greatly improved breakdown field strength and D_m-D_r in B1B composites may lead to simultaneous improvements in discharge energy density and energy efficiency.

In the above analysis, the Sawyer-Tower circuit was used to test the discharge energy density and charge-discharge efficiency of the sandwich structure composite at room temperature. The discharge energy density and energy efficiency of the sandwich-structured composites can be calculated from the D-E curves, and the results are shown in Figure 8b.

$$\eta = \frac{U_d}{U} \tag{3}$$

In Formula (3), U_d is the released energy density of the material, and U is the stored energy density of the material [41]. For a more intuitive comparison, Figure 8b presents detailed data on the discharge energy density, charge energy density, and charge-discharge energy efficiency under the breakdown electric field for the four sandwich-structured composites. Although the maximum electrical displacement of B3B is larger than that of B2B, the discharge energy density of B2B is higher due to the higher breakdown field strength of B2B. At the same time, the higher D_r of B2B also leads to a significant drop in its energy efficiency. B1B exhibited the highest discharge energy density of 8.9 J/cm^3 at a breakdown field of 310 kV/mm, an improvement over the energy storage performance data of PVDF obtained under the same conditions, which is shown in Figure S4. Although the energy efficiency of the composite gradually decreased with the electric field due to the ferroelectricity of PVDF and $Ba_{0.6}Sr_{0.4}TiO_3$ NPs, the energy efficiency of B1B remained at 76% at an electric field of 310 kV/mm. Taking the discharge energy density and energy efficiency into consideration, B1B exhibits the best energy storage performance among a series of sandwich-structured composite dielectrics.

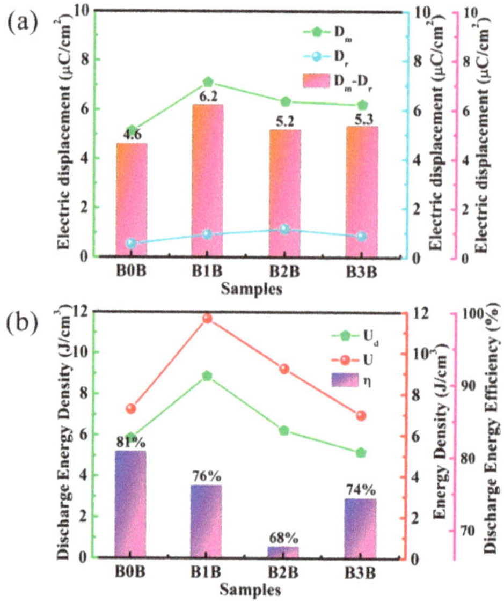

Figure 8. (**a**) Maximum electric displacement (D_m), residual electrical displacement (D_r), and the difference between residual electrical displacement and maximum electric displacement (D_m-D_r) as a function of $Ba_{0.6}Sr_{0.4}TiO_3$ NP content in the middle layer of the sandwich-structured nanocomposites. (**b**) Maximum discharged energy density (U_d), charged energy density (U), and charged-discharge energy efficiency (η) as a function of the different sandwich-structured nanocomposites with varied $Ba_{0.6}Sr_{0.4}TiO_3$ NP contents in the middle layer.

4. Conclusions

In this study, PVDF-based sandwich structure nanocomposites were prepared by a layer-by-layer solution casting method using PVDF as the matrix. In the sandwich-structured nanocomposites, 5 wt.% PVDF/BNNS was used as the outer layer of the three-layer structure, and 0 vol.%, 1 vol.%, 2 vol.%, and 3 vol.% PVDF/$Ba_{0.6}Sr_{0.4}TiO_3$ NPs were used as the middle layer, respectively. The dispersion of BNNS by ultrasound and the coating of $Ba_{0.6}Sr_{0.4}TiO_3$ NPs by dopamine promote the uniform distribution of nanofillers in the polymer matrix. SEM analysis showed that the layers of the sandwich-structured nanocomposites are tightly bonded to the other layers. EDS showed that the distribution of layers is very clear and the BNNS and $Ba_{0.6}Sr_{0.4}TiO_3$ NPs are well dispersed. XRD analysis showed that the composites were mainly composed of non-polar α-phase and γ-phase, which was conducive to the rotation of the electric dipoles of the composites during the discharge process. The breakdown field strength of the sandwich structure composite with 1 vol.% PVDF/$Ba_{0.6}Sr_{0.4}TiO_3$ in the middle layer reaches 310 Kv/mm. D-E curve analysis shows that the maximum discharge energy density of the sandwich-structured composite is 8.9, and the maximum discharge energy efficiency is 76%.

Supplementary Materials: The following supporting information can be downloaded at: https://www.mdpi.com/article/10.3390/polym15173642/s1, Figure S1. Schematic diagram of fabrication of $Ba_{0.6}Sr_{0.4}TiO_3$ nanoparticles; Figure S2. The distribution of diameter of $Ba_{0.6}Sr_{0.4}TiO_3$ nanoparticles; Figure S3. TEM images of $Ba_{0.6}Sr_{0.4}TiO_3$@DA NPs; Figure S4. The energy storage performance data of PVDF obtained under the same conditions.

Author Contributions: Conceptualization, Z.Y. and Z.W.; methodology, Z.Y. and Z.W.; software, Z.W.; validation, Z.Y. and Z.W.; formal analysis, Z.Y. and Z.W.; investigation, Z.Y.; resources, Z.W.; data curation, Z.Y., D.W. and Y.X.; writing—original draft preparation, Z.Y.; writing—review and editing, Z.Y., Z.W., D.W. and Y.X.; supervision, Z.W.; project administration, Z.Y., Z.W., D.W. and Y.X.; funding acquisition, Z.W. All authors have read and agreed to the published version of the manuscript.

Funding: The present work was supported by the National Natural Science Foundation of China (52102139), the National Key Research and Development Program of Shaanxi Province (2021GY-224), the Key Laboratory of Engineering Dielectrics and Its Application (Harbin University of Science and Technology), the Ministry of Education (KFM202106), and the Graduate Innovation Fund of Shaanxi University of Science and Technology.

Institutional Review Board Statement: Not applicable.

Informed Consent Statement: Not applicable.

Data Availability Statement: The data presented in this study are available on request from the corresponding author.

Conflicts of Interest: The authors declare that they have no conflict of interest.

References

1. Chu, B.J.; Zhou, X.; Ren, K.L.; Neese, B.; Lin, M.R.; Wang, Q.; Bauer, F.; Zhang, Q.M. A dielectric polymer with high electric energy density and fast discharge speed. *Science* **2006**, *313*, 334–336. [CrossRef] [PubMed]
2. Shen, X.; Zheng, Q.; Kim, J.-K. Rational design of two-dimensional nanofillers for polymer nanocomposites toward multifunctional applications. *Prog. Mater. Sci.* **2021**, *115*, 100708. [CrossRef]
3. Chen, J.; Wang, Y.; Yuan, Q.; Xu, X.; Niu, Y.; Wang, Q.; Wang, H. Multilayered ferroelectric polymer films incorporating low-dielectric-constant components for concurrent enhancement of energy density and charge-discharge efficiency. *Nano Energy* **2018**, *54*, 288–296. [CrossRef]
4. Chi, Q.; Zhou, Y.; Feng, Y.; Cui, Y.; Zhang, Y.; Zhang, T.; Chen, Q. Excellent energy storage performance of polyetherimide filled by oriented nano fibers with optimized diameters. *Mater. Today Energy* **2020**, *18*, 100516. [CrossRef]
5. Martins, P.; Lopes, A.C.; Lanceros-Mendez, S. Electroactive phases of poly(vinylidene fluoride): Determination, processing and applications. *Prog. Polym. Sci.* **2014**, *39*, 683–706. [CrossRef]
6. Feng, Q.-K.; Zhong, S.-L.; Pei, J.-Y.; Zhao, Y.; Zhang, D.-L.; Liu, D.-F.; Zhang, Y.-X.; Dang, Z.-M. Recent Progress and Future Prospects on All-Organic Polymer Dielectrics for Energy Storage Capacitors. *Chem. Rev.* **2022**, *122*, 3820–3878. [CrossRef]
7. Yi, Z.; Wang, Z.; Nian, W.; Wang, T.; Chen, H.; Cheng, Z. High Energy Storage Density of Sandwich-Structured Na0.5Bi0.5TiO3/PVDF Nanocomposites Enhanced by Optimizing the Dimensions of Fillers. *ACS Appl. Energy Mater.* **2021**, *4*, 13528–13537. [CrossRef]
8. Jiang, J.; Shen, Z.; Cai, X.; Qian, J.; Dan, Z.; Lin, Y.; Liu, B.; Nan, C.-W.; Chen, L.; Shen, Y. Polymer Nanocomposites with Interpenetrating Gradient Structure Exhibiting Ultrahigh Discharge Efficiency and Energy Density. *Adv. Energy Mater.* **2019**, *9*, 1803411. [CrossRef]
9. Khanchaitit, P.; Han, K.; Gadinski, M.R.; Li, Q.; Wang, Q. Ferroelectric polymer networks with high energy density and improved discharged efficiency for dielectric energy storage. *Nat. Commun.* **2013**, *4*, 2845. [CrossRef]
10. Sun, S.; Shi, Z.; Liang, L.; Li, T.; Zhang, S.; Xu, W.; Han, M.; Zhang, M. Simultaneous Realization of Significantly Enhanced Breakdown Strength and Moderately Enhanced Permittivity in Layered PMMA/P(VDF-HFP) Nanocomposites via Inserting an Al_2O_3/P(VDF-HFP) Layer. *J. Phys. Chem. C* **2021**, *125*, 22379–22387. [CrossRef]
11. Diao, C.; Liu, H.; Zheng, H.; Yao, Z.; Iqbal, J.; Cao, M.; Hao, H. Enhanced energy storage properties of $BaTiO_3$ thin films by $Ba_{0.4}Sr_{0.6}TiO_3$ layers modulation. *J. Alloys Compd.* **2018**, *765*, 362–368. [CrossRef]
12. Chen, Q.; Shen, Y.; Zhang, S.H.; Zhang, Q.M. Polymer-Based Dielectrics with High Energy Storage Density. *Annu. Rev. Mater. Res.* **2015**, *45*, 433–458. [CrossRef]
13. Pan, Z.; Yao, L.; Zhai, J.; Yao, X.; Chen, H. Interfacial Coupling Effect in Organic/Inorganic Nanocomposites with High Energy Density. *Adv. Mater.* **2018**, *30*, 1705662. [CrossRef] [PubMed]
14. Hu, J.; Zhang, S.; Tang, B. 2D filler-reinforced polymer nanocomposite dielectrics for high-k dielectric and energy storage applications. *Energy Storage Mater.* **2021**, *34*, 260–281. [CrossRef]
15. Li, H.; Ren, L.; Zhou, Y.; Yao, B.; Wang, Q. Recent progress in polymer dielectrics containing boron nitride nanosheets for high energy density capacitors. *High Volt.* **2020**, *5*, 365–376. [CrossRef]
16. Hu, H.; Zhang, F.; Luo, S.; Chang, W.; Yue, J.; Wang, C.-H. Recent advances in rational design of polymer nanocomposite dielectrics for energy storage. *Nano Energy* **2020**, *74*, 104844. [CrossRef]
17. Cui, Y.; Zhang, T.; Feng, Y.; Zhang, C.; Chi, Q.; Zhang, Y.; Chen, Q.; Wang, X.; Lei, Q. Excellent energy storage density and efficiency in blend polymer-based composites by design of core-shell structured inorganic fibers and sandwich structured films. *Compos. Part B Eng.* **2019**, *177*, 107429. [CrossRef]

18. Li, Q.; Zhang, G.Z.; Liu, F.H.; Han, K.; Gadinski, M.R.; Xiong, C.X.; Wang, Q. Solution-processed ferroelectric terpolymer nanocomposites with high breakdown strength and energy density utilizing boron nitride nanosheets. *Energy Environ. Sci.* **2015**, *8*, 922–931. [CrossRef]
19. Pan, Z.; Zhai, J.; Shen, B. Multilayer hierarchical interfaces with high energy density in polymer nanocomposites composed of BaTiO3@TiO2@Al2O3 nanofibers. *J. Mater. Chem. A* **2017**, *5*, 15217–15226. [CrossRef]
20. Li, Y.; Zhou, Y.; Zhu, Y.; Cheng, S.; Yuan, C.; Hu, J.; He, J.; Li, Q. Polymer nanocomposites with high energy density and improved charge-discharge efficiency utilizing hierarchically-structured nanofillers. *J. Mater. Chem. A* **2020**, *8*, 6576–6585. [CrossRef]
21. Bai, H.; Zhu, K.; Wang, Z.; Shen, B.; Zhai, J. 2D Fillers Highly Boost the Discharge Energy Density of Polymer-Based Nanocomposites with Trilayered Architecture. *Adv. Funct. Mater.* **2021**, *31*, 2102646. [CrossRef]
22. Pan, Z.; Ding, Q.; Yao, L.; Huang, S.; Xing, S.; Liu, J.; Chen, J.; Zhai, J. Simultaneously enhanced discharge energy density and efficiency in nanocomposite film capacitors utilizing two-dimensional NaNbO3@Al2O3 platelets. *Nanoscale* **2019**, *11*, 10546–10554. [CrossRef] [PubMed]
23. Sun, L.; Shi, Z.; He, B.; Wang, H.; Liu, S.; Huang, M.; Shi, J.; Dastan, D.; Wang, H. Asymmetric Trilayer All-Polymer Dielectric Composites with Simultaneous High Efficiency and High Energy Density: A Novel Design Targeting for Advanced Energy Storage Capacitors. *Adv. Funct. Mater.* **2021**, *31*, 2100280. [CrossRef]
24. Cheng, Y.; Pan, Z.; Bai, H.; Chen, H.; Yao, L.; Ding, X.; Shi, S.; Liu, J.; Xie, Z.; Xu, J.; et al. Two-Dimensional Fillers Induced Superior Electrostatic Energy Storage Performance in Trilayered Architecture Nanocomposites. *ACS Appl. Mater. Interfaces* **2022**, *14*, 8448–8457. [CrossRef] [PubMed]
25. Chi, Q.; Ma, T.; Zhang, Y.; Chen, Q.; Zhang, C.; Cui, Y.; Zhang, T.; Lin, J.; Wang, X.; Lei, Q. Excellent Energy Storage of Sandwich-Structured PVDF-Based Composite at Low Electric Field by Introduction of the Hybrid $CoFe_2O_4$@BZT-BCT Nanofibers. *ACS Sustain. Chem. Eng.* **2018**, *6*, 403–412. [CrossRef]
26. Sun, Q.; Wang, J.; Sun, H.; He, L.; Zhang, L.; Mao, P.; Zhang, X.; Kang, F.; Wang, Z.; Kang, R.; et al. Simultaneously enhanced energy density and discharge efficiency of layer-structured nanocomposites by reasonably designing dielectric differences between $BaTiO_3$@SiO_2/PVDF layers and BNNSs/PVDF-PMMA layers. *Compos. Part A Appl. Sci. Manuf.* **2021**, *149*, 106546. [CrossRef]
27. Hu, P.H.; Shen, Y.; Guan, Y.H.; Zhang, X.H.; Lin, Y.H.; Zhang, Q.M.; Nan, C.W. Topological-Structure Modulated Polymer Nanocomposites Exhibiting Highly Enhanced Dielectric Strength and Energy Density. *Adv. Funct. Mater.* **2014**, *24*, 3172–3178. [CrossRef]
28. Li, Z.; Shen, Z.; Yang, X.; Zhu, X.; Zhou, Y.; Dong, L.; Xiong, C.; Wang, Q. Ultrahigh charge-discharge efficiency and enhanced energy density of the sandwiched polymer nanocomposites with poly(methyl methacrylate) layer. *Compos. Sci. Technol.* **2021**, *202*, 108591. [CrossRef]
29. Xu, W.; Liu, J.; Chen, T.; Jiang, X.; Qian, X.; Zhang, Y.; Jiang, Z.; Zhang, Y. Bioinspired Polymer Nanocomposites Exhibit Giant Energy Density and High Efficiency at High Temperature. *Small* **2019**, *15*, 1901582. [CrossRef]
30. Li, Z.; Liu, F.; Li, H.; Ren, L.; Dong, L.; Xiong, C.; Wang, Q. Largely enhanced energy storage performance of sandwich-structured polymer nanocomposites with synergistic inorganic nanowires. *Ceram. Int.* **2019**, *45*, 8216–8221. [CrossRef]
31. Zhu, D.; Zhao, J.; Yang, Z.; Guo, H.; Gao, L. Graphene Oxide/Polyimide Composites with High Energy Storage Density Based on Multilayer Structure. *Chem. J. Chin. Univ. Chin.* **2021**, *42*, 2694–2700. [CrossRef]
32. Wang, Y.; Wang, L.; Yuan, Q.; Niu, Y.; Chen, J.; Wang, Q.; Wang, H. Ultrahigh electric displacement and energy density in gradient layer-structured BaTiO3/PVDF nanocomposites with an interfacial barrier effect. *J. Mater. Chem. A* **2017**, *5*, 10849–10855. [CrossRef]
33. Huang, Y.H.; Wang, J.J.; Yang, T.N.; Wu, Y.J.; Chen, X.M.; Chen, L.Q. A thermodynamic potential, energy storage performances, and electrocaloric effects of Ba1-xSrxTiO3 single crystals. *Appl. Phys. Lett.* **2018**, *112*, 102901. [CrossRef]
34. Chen, S.; Xu, R.; Liu, J.; Zou, X.; Qiu, L.; Kang, F.; Liu, B.; Cheng, H.-M. Simultaneous Production and Functionalization of Boron Nitride Nanosheets by Sugar-Assisted Mechanochemical Exfoliation. *Adv. Mater.* **2019**, *31*, 1804810. [CrossRef] [PubMed]
35. Feng, M.; Feng, Y.; Zhang, T.; Li, J.; Chen, Q.; Chi, Q.; Lei, Q. Recent Advances in Multilayer-Structure Dielectrics for Energy Storage Application. *Adv. Sci.* **2021**, *8*, 2102221. [CrossRef]
36. Chen, J.; Liu, J.; Cai, L.; Wang, C.; Liu, L.; Yang, Q.; Xiong, C. Incorporation of elaborately Synthesized BNNSs by a mild mechanical stirring process for the concurrent enhancement of thermal conductivity and dielectric breakdown strength of PVDF. *Compos. Sci. Technol.* **2020**, *200*, 108381. [CrossRef]
37. Marwat, M.A.; Ma, W.; Fan, P.; Elahi, H.; Samart, C.; Nan, B.; Tan, H.; Salamon, D.; Ye, B.; Zhang, H. Ultrahigh energy density and thermal stability in sandwich-structured nanocomposites with dopamine@Ag@BaTiO3. *Energy Storage Mater.* **2020**, *31*, 492–504. [CrossRef]
38. Nakafuku, C.; Yasuniwa, M. Melting and Crystallization of Poly(vinylidene fluoride) in the Blend with Poly(methyl methacrylate) under High Pressure. *Polym. J.* **1987**, *19*, 845–853. [CrossRef]
39. Zhang, Y.; Zhang, C.; Feng, Y.; Zhang, T.; Chen, Q.; Chi, Q.; Liu, L.; Wang, X.; Lei, Q. Energy storage enhancement of P(VDF-TrFE-CFE)-based composites with double-shell structured BZCT nanofibers of parallel and orthogonal configurations. *Nano Energy* **2019**, *66*, 104195. [CrossRef]

40. Zhu, L.; Zhang, Y.; Xu, W.; Zhu, X.; Niu, S.; Zhang, Y.; Jiang, Z. Crosslinked polyetherimide nanocomposites with superior energy storage achieved via trace Al_2O_3 nanoparticles. *Compos. Sci. Technol.* **2022**, *223*, 109421. [CrossRef]
41. Wang, C.; He, G.; Chen, S.; Luo, H.; Yang, Y.; Zhang, D. Achieving high breakdown strength and energy density in all-organic sandwich-structured dielectrics by introducing polyacrylate elastomers. *J. Mater. Chem. A* **2022**, *10*, 9103–9113. [CrossRef]

polymers

MDPI

Article

Applying a Green Solvent with Microwave, Ultrasound, and Soxhlet Extraction Techniques to Quantify the Slip Additive *cis*-1,3-Docosenamide and Nine Oxidative Degradation Byproducts in Polypropylene Samples

Hernández Fernández Joaquin [1,2,3,*], Pérez Mendoza Jaime [4] and Ortega-Toro Rodrigo [5]

[1] Chemistry Program, Department of Natural and Exact Sciences, San Pablo Campus, University of Cartagena, Cartagena 130015, Colombia
[2] Chemical Engineering Program, School of Engineering, Universidad Tecnológica de Bolivar, Parque Industrial y Tecnológico Carlos Vélez Pombo Km 1 Vía Turbaco, Cartagena 130001, Colombia
[3] Department of Natural and Exact Science, Universidad de la Costa, Barranquilla 080002, Colombia
[4] Complex Fluid Engineering and Food Rheology Research Group (IFCRA), Food Engineering Department, Universidad de Cartagena, Cartagena de Indias 130015, Colombia; jperezm@unicartagena.edu.co
[5] Food Packaging and Shelf-Life Research Group (FP&SL), Food Engineering Department, Universidad de Cartagena, Cartagena de Indias 130015, Colombia; rortegap1@unicartagena.edu.co
* Correspondence: jhernandezf@unicartagena.edu.co; Tel.: +57-301-562-4990

Abstract: Erucamide is used as an important slip agent for polymers. However, erucamide can degrade during processing and long-term storage, forming various oxidation products. These degradation products can affect the recovery rates of erucamide. In this study, investigated different solid–liquid extraction methods (Soxhlet, microwave, and ultrasound) and used gas chromatography with mass spectrometry (GC-MS) to quantify erucamide and its degradation byproducts in polypropylene (PP). A multivariable experiment was designed, and a mixed-effect approach was used to analyze the results. Various extraction variables were examined, such as temperature, time, solvents, and PP pretreatments. Using a mixed-effect model with a Kenward–Roger approximation, an R2 of the model of 97% and p values of 0.168, 0.000, and 0.000 were obtained for the technical, solvent, and type of PP pretreatment variables, respectively. The highest average recoveries of erucamide were found with the microwave technique and were 96.4% using dichloromethane, 94.57% using cyclohexane, and 93.05% using limonene. With ultrasound, recoveries ranged between 85 and 92% for dichloromethane and limonene. In addition, it was observed that the extraction method had better recovery results in ground PP than in films and in pellets. Nine oxidative degradation byproducts of erucamide were identified and semi-quantified by GC-MS. The reaction mechanisms for forming each byproduct were proposed. The byproducts that experienced a higher rate of degradation of erucamide were erucamide with a hydroxyl group at position one and 12-amino-6-12-oxo-dodecanoic acid, showing more prominent peaks using the Soxhlet method with cyclohexane and dichloromethane as solvents and polypropylene (PP) films as the type of material used.

Keywords: erucamide; extraction; microwave; Soxhlet; ultrasound; cyclohexane; dichloromethane; limonene; GC-MS

Citation: Joaquin, H.F.; Jaime, P.M.; Rodrigo, O.-T. Applying a Green Solvent with Microwave, Ultrasound, and Soxhlet Extraction Techniques to Quantify the Slip Additive *cis*-1,3-Docosenamide and Nine Oxidative Degradation Byproducts in Polypropylene Samples. *Polymers* **2023**, *15*, 3457. https://doi.org/10.3390/polym15163457

Academic Editors: Hui Zhao, Wei Wu and Bin Wang

Received: 10 July 2023
Revised: 15 August 2023
Accepted: 16 August 2023
Published: 18 August 2023

1. Introduction

There is a tremendous demand for polypropylene films with different characteristics, such as multilayer films, materials for controlled atmosphere, and barrier materials, especially in the packaging industry [1–5]. In their various formulations, these films must meet intrinsic and extrinsic requirements for their preparation, handling, and application and for their final use [6–10]. There are two fundamental functions that these films must fulfill. In the first place, they must be easily manipulated by the corresponding machinery in the packaging or packaging lines without the occurrence of phenomena such as electrification or adhesion between the film and the product, between films, or between the film and any part of the machine [11–14]. This implies that they must have good sliding and anti-blocking properties. To achieve this, during the industrial manufacture of the films and their incorporation into the raw materials, additives known as slip and anti-block agents are added. Among the commonly used slip additives in polypropylene (PP) are long-chain fatty acid amides [15–18]. These additives help to reduce friction and facilitate the sliding of the films during processing and handling, avoiding problems such as blocking or unwanted adhesions [19].

Erucamide is an amide belonging to the monounsaturated fatty acids group, characterized by a hydrocarbon chain composed of 22 carbon atoms in its chemical structure [15,20]. Erucamide is thermally stable at a relatively higher temperature than other slip agents, such as oleamide [15]. During the extrusion and molding of PP, erucamide dissolves in the polymer. Then erucamide crystallizes in the polymer to form a lubricating layer on the surface of the solidified polymer [15]. During the erucamide dosing process, erucamide may be added along with anti-blocking agents such as silica or talc; in such cases, erucamide is absorbed onto the surface of the anti-blocking agent [15]. As a combination, erucamide, and silica reduce the coefficient of friction of PP more effectively than either alone [1]. As a general rule, it is essential to note that the most effective slip agent is only sometimes the best anti-blocking agent. In this context, erucamide is a slip and anti-block additive that is preferred in PP-producing industries due to its higher melting point and higher heat resistance [20]. These properties make erucamide more suitable for applications where higher thermal stability and resistance to surface blocking are required than PP films [21–24].

Traditionally extraction of erucamide from polymers has been performed by Soxhlet extraction, ultrasonic-assisted extraction (UAE) [25], supercritical fluid extraction (SFE) [26], and microwave-assisted extraction (MAE) [26,27]. Although Soxhlet extraction achieves good efficiency, it is slow [27]. On the other hand, ultrasonic and microwave extraction are environmentally friendly techniques that offer several advantages over conventional methods. These advantages include reduced extraction time, lower solvent use, and higher yield of the chemical of interest. These ecological techniques allow more efficient and sustainable extractions. Some of these techniques have been used individually to extract erucamide. Still, these extraction techniques have not been studied simultaneously with green solvents, and the effect of each extraction technique on the oxidative degradation of erucamide during its extraction from the PP matrix has not yet been investigated [28–32].

Selection of the extraction solvent is crucial to achieving complete extraction, preventing it from reacting with erucamide, preventing it from completely solubilizing the polymer, determining the duration of extraction in each of the extraction techniques, and swelling the polymer. And improve the miscibility and diffusion within the polymer. This is important because as the polymer swells, the solvent penetrates the intermolecular spaces of the polymer, separating the polymer chains and allowing the erucamide to be extracted to dissolve more quickly in the solvent [33]. Commonly, solvents such as cyclohexane and dichloromethane are used in solid–liquid extraction [34,35]. However, cyclohexane has occupational and environmental disadvantages due to its rapid absorption in the human body [36]. On the other hand, dichloromethane has been reported to have harmful health and environmental effects, including the risk of diseases such as brain and liver cancer and DNA damage [37–41]. Due to the potential dangers associated with cyclohexane and dichloromethane, safer and more environmentally friendly alternatives are sought, such as green solvents [37–44]. One of the leading candidates as a green solvent is limonene, which belongs to the chemical family of monoterpenes (Figure 1) and is a natural compound.

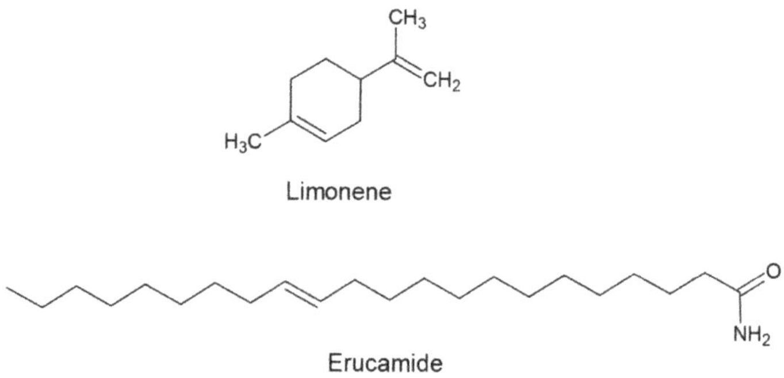

Figure 1. Chemical structure of erucamide and limonene.

Limonene is known as an ecological solvent and, therefore, an excellent substitute for hazardous solvents in multiple aspects [45–48]. Limonene is a colorless liquid that has a high capacity to dissolve both polar and nonpolar substances, making it ideal for use as a solvent in industrial applications. Limonene is biodegradable and non-toxic, making it safer and more environmentally friendly than many synthetic solvents. It is beneficial as a solvent in the cleaning industry and is used in household and commercial cleaning products. Limonene as a solvent has also been explored in other applications, including pharmaceuticals, food and beverages, and extracting organic compounds from plants. Limonene as a biosolvent is considered a sustainable and efficient alternative to traditional solvents, making it an increasingly popular option in the industry [49–53]. Based on the latest EU OEL data, no specific occupational exposure limit is set for limonene. However, the MAK value (Maximale Arbeitsplatzkonzentration) of 28 mg/m^3 is reported, which refers to the occupational exposure limit established for limonene by the Hazardous Substances Committee of the German Research Foundation for Occupational Safety and Health (Deutsche Forschungsgemeinschaft (DFG)) [54]. This value can be considered relatively "good" compared to the values reported for cyclohexane and dichloromethane. It is important to note that MAK values are specific to Germany and may vary in other countries or regions. However, the MAK value of 28 mg/m^3 can be a valuable reference to assess and monitor exposure to limonene in the workplace [54]. Limonene is an excellent biosolvent due to its chemical and physical properties, such as its dipole moment, electronegativity, and reactivity. These physicochemical properties should allow a very low reactivity and degradation of erucamide. This investigation evaluated the effect of dichloromethane, cyclohexane, and limonene on the degradation of erucamide.

In our research, we performed microwave-assisted solvent extraction, ultrasonic-assisted extraction, and Soxhlet extraction based on a sensitive microextraction method, flame ionization detector gas chromatography, and mass spectrometry. In each extraction technique, the erucamide recovery percentage was evaluated, and the extent to which each method affects the formation of erucamide oxidation or degradation byproducts was determined. For the degradation byproducts formed, we propose respective reaction mechanisms that allow us to understand how erucamide degradation occurs. We carried out three previous PP treatments (ground PP, PP granules, and PP films) using three different solvents (dichloromethane, cyclohexane, and limonene) and determined how each of these procedures affects the generation of nine additive degradation byproducts. The performance of the multiple variables was evaluated through multivariate statistical analysis, allowing the development of a multiple regression model that allows a comprehensive understanding and control of most factors present in the quantification of erucamide in polypropylene. This research should support other researchers, polypropylene-producing industries, film manufacturers, erucamide-producing industries, and chemical regulators interested in quantifying erucamide from a polymeric matrix and knowing its performance as a slip and anti-blocking agent.

2. Materials and Methods

2.1. Materials and Reagents

The working cis-1,3-docoseneamide (erucamide) was supplied by Cymit Quimica Croda Universal (4014 Walnut Pond Drive Houston, TX 77059 281-282-0022 Crompton (Witco) Corporation) (erucamide has a white coloration; its iodine value ranged from 75 to 82. The result of the acid number was 0.1 KOH mg/g, and the pour point ranged from 78 to 81 °C. The moisture content obtained was 0.4% max. The certified purity was 99% min). N-tetradecanamide (Alpha Aesar, Karlsruhe, Germany) was used as internal standard. Limonene (HPLC grade) was obtained from Scharlab (Barcelona, Spain). Hydrogen 99.9999% was from Linde (Cartagena, Colombia), nitrogen 99.9999% was from Linde (Cartagena, Colombia), cyclohexane 99.5% was from Panreac (Barcelona, Spain), and dichloromethane 99.99% was from Sigma Aldrich (Bangalore, India).

2.2. Instrumentation

An Agilent 6890 gas chromatograph (GC) (Agilent Technologies, Wilmington, CA, USA) with a mass detector (MS) was used to measure the samples. The MS detector was heated to 230 °C. An Agilent J&W VF-5 ms column (5% phenyl and 95% dimethylpolysiloxane) with 30 m × 0.25 mm i.d., with a diameter of 0.25 m, was used. The oven heating cycle started at 200 °C for 4 min, rose to 280 °C at a rate of 10 °C min^{-1}, and remained there for 7 min. Helium, at 1.0 mL min^{-1}, was the carrier gas (99.996%). The injection system was in splitless mode. One µL of the sample was injected. The GC-MS apparatus was operated and the data were processed with Chemstation software. With these chromatographic parameters, it was possible to obtain a retention time (T_r) of 7.5 min for erucamide.

2.2.1. Preparation of Erucamide Calibration Standards and PP Samples with Erucamide

Preparation of the Curve for Chromatograph Calibration

A stock solution of erucamide at 10,000 ppm was prepared (10,000 mg of erucamide is weighed and 1 L of limonene was added). In another vessel, an internal standard solution of n-tetradecanamide at 10,000 ppm was prepared. Using the erucamide stock solution and the internal standard solution, six calibration standards were developed with concentrations of 5000, 3000, 2000, 1500, 1000, and 500 ppm erucamide (see Figure 2).

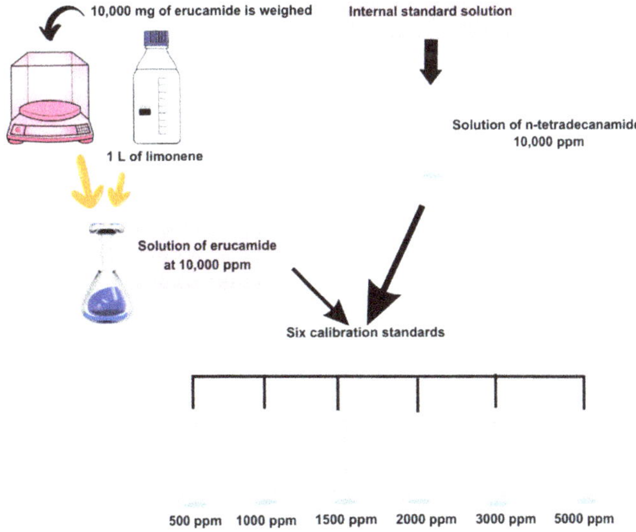

Figure 2. Calibration curve preparation.

2.3. Preparation of PP Samples with Different Concentrations of Erucamide

The PP and erucamide samples were prepared as follows: (1) Individually, 0.0, 0.5, 1, 1.5, 1.5, 2, 3, and 5 g of erucamide were weighed. (2) Each amount of erucamide was mixed with 1 kg of virgin PP resin. (3) Each mixture was premixed at 800 rpm x 7 min and using a standard Prodex Henschel 115JSS mixer (Federal Equipment Company, NJ, EE.UU). Each sample was then extruded in a Welex-200 24 extruder (KD Capital Equipment, LLC, CA, USA). The extruder operated with five temperature zones along the entire extrusion path. The temperatures were 190, 195, 200, 210, 210, and 220 °C. In this way, uniform mixing was achieved. At the outlet of the extruder, a PP-Erucamide melt mix was produced. For each melt type, 20 g of melt was fed to a CARVER 3895 hot press (SPECTRA SERVICES, INC., NY 14519, EE.UU). In this CARVER machine, the samples were compressed to form films of 300 mm diameter and ≈100 μm thickness. The films obtained in the experiment were identified as PP (0 ppm erucamide), PP2 (500 ppm erucamide), PP3 (1000 ppm erucamide), PP4 (1500 ppm erucamide), PP5 (2000 ppm erucamide), PP6 (3000 ppm erucamide), and PP7 (5000 ppm erucamide) (Figure 3).

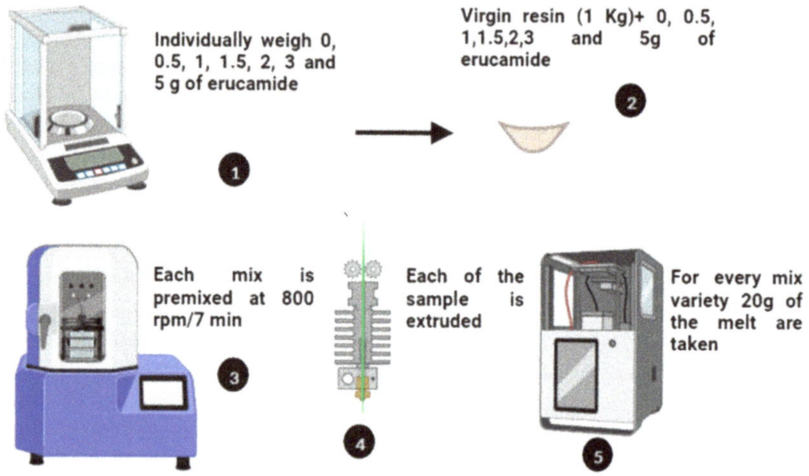

Figure 3. Preparation of PP samples with different concentrations of erucamide.

2.4. Extraction of Erucamide from PP Samples

We obtained PP samples in the form of pellets, films, and grinds, to which erucamide was added. Three different extraction solvents were used, the first one being cyclohexane. The second option (which proved to be more efficient) was dichloromethane. The third option was limonene. Using each of the solvents, 3 different extraction methods were tested: Soxhlet, ultrasound (conventional laboratory sonic bath), and microwave oven (high-power, programmable laboratory microwave oven). Figure 4 shows the outline of the methodology followed in the investigation.

For the ultrasonic bath, 3 g of PP was added to a 30 mL vial, and then 20.0 mL of the internal standard solution was added using a 5.0 mL micropipette. For each assay, 5 replicates were performed. For three hours, the sonication procedure was carried out in an ultrasonic bath. The temperature was kept under control during the sonication process up to a maximum of 50 °C. After the sonication process was finished, the vials were taken out of the ultrasonic bath and left outside for a duration of 10 min. Disposable PTFE syringe filters were used to filter the extracted erucamide sample solutions. Crushed and pelleted PP and films were extracted for 90 and 60 min in the ultrasonic bath, with the solution agitated for at least 30 s every 10 min.

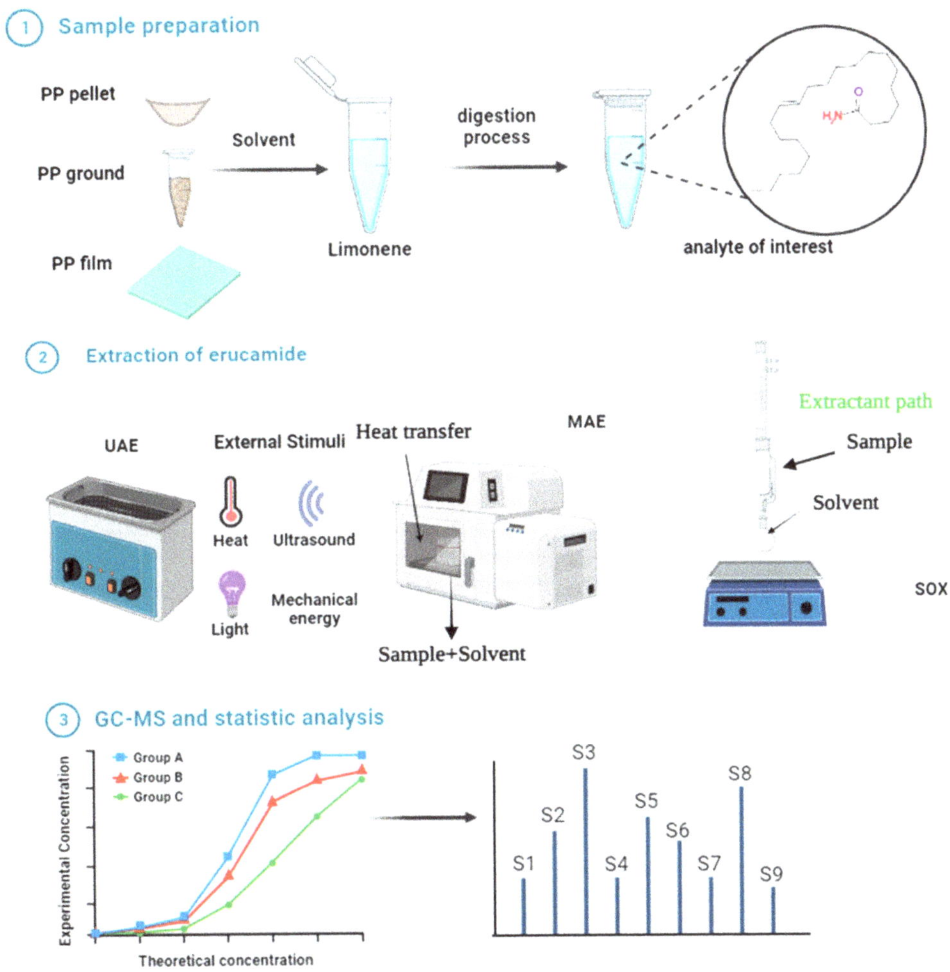

Figure 4. Erucamide extraction by microwave, Soxhlet, and ultrasound and quantification by GC-MS.

For the microwave oven, five grams of PP resin was extracted using cyclohexane, dichloromethane, and limonene. It was found that only heating the solution in the microwave oven for 25 min at 50% power was required to extract the slip agent, this process was also performed by heating for 45 min at 25% power with stirring every 5 min. Six different extractions were performed with the resin, pellets, and ground PP, and the average results for ultrasonic, Soxhlet, and microwave extraction are shown in Table 1. The microwave oven provided a very rapid means of extracting the erucamide from the crushed resin. The ultrasonic bath provided an economical and relatively fast way to extract the additives. The Soxhlet method of extraction with these polypropylene resins took at least 7 h to extract most of the additives. In our case, Soxhlet extraction was carried out for 1440 and 720 min and possibly took more than 24 h to fully recover the additive.

Table 1. Erucamide recovery percentage according to the experimental design.

Technique	Time (min)	Temperature (°C)	Solvent	Form of PP	Erucamide Recovery %					
					PP2	PP3	PP4	PP5	PP6	PP7
Microwave	45	117	Cyclohexane	Pellets	85.40	82.82	85.23	87.25	86.85	86.84
Microwave	45	117	Cyclohexane	Films	87.5	92.65	94.67	92.25	93.17	92.45
Microwave	45	117	Cyclohexane	PP-ground	91.78	94.50	96.67	95.25	95.33	93.88
Microwave	25	117	Cyclohexane	Pellets	76.50	76.35	72.30	78.83	83.68	83.54
Microwave	25	117	Cyclohexane	Films	77.40	81.70	79.67	85.95	81.50	84.20
Microwave	25	117	Cyclohexane	PP-ground	87.39	92.30	93.17	93.68	91.93	92.76
Soxhlet	1440	90	Cyclohexane	Pellets	65.80	68.95	67.83	65.95	70.49	71.54
Soxhlet	1440	90	Cyclohexane	Films	74.10	70.85	72.00	72.69	73.91	78.59
Soxhlet	1440	90	Cyclohexane	PP-ground	79.00	78.33	78.88	78.20	79.5	78.73
Soxhlet	720	90	Cyclohexane	Pellets	45.50	39.45	38.70	48.45	53.78	51.42
Soxhlet	720	90	Cyclohexane	Films	52.15	56.95	49.57	56.88	53.78	51.42
Soxhlet	720	90	Cyclohexane	PP-ground	65.80	66.40	68.83	65.20	67.11	69.55
Ultrasound	90	50	Cyclohexane	Pellets	83.80	81.14	80.8	84.69	82.07	84.17
Ultrasound	90	50	Cyclohexane	Films	86.00	82.92	80.05	83.45	79.83	89.61
Ultrasound	90	50	Cyclohexane	PP-ground	88.56	88.52	82.84	87.62	85.55	92.46
Ultrasound	60	50	Cyclohexane	Pellets	77.50	77.37	75.23	79.47	77.28	78.7
Ultrasound	60	50	Cyclohexane	Films	80.40	78.26	74.86	78.85	76.16	78.3
Ultrasound	60	50	Cyclohexane	PP-ground	84.28	81.00	80.25	82.36	82.54	81.4
Microwave	45	117	Dichloromethane	Pellets	89.14	90.26	92.39	91.44	91.03	90.67
Microwave	45	117	Dichloromethane	Films	94.92	94.28	92.79	93.68	91.65	93.4
Microwave	45	117	Dichloromethane	PP-ground	96.44	96.54	97.29	94.94	94.74	98.35
Microwave	25	117	Dichloromethane	Pellets	86.07	87.38	87.69	86.84	86.76	88.51
Microwave	25	117	Dichloromethane	Films	89.90	84.75	86.37	87.97	87.08	89.88
Microwave	25	117	Dichloromethane	PP-ground	92.60	90.52	93.48	92.37	93.19	93.92
Soxhlet	1440	90	Dichloromethane	Pellets	78.60	72.60	74.53	79.18	71.98	76.93
Soxhlet	1440	90	Dichloromethane	Films	78.60	77.85	87.33	84.20	81.98	84.21
Soxhlet	1440	90	Dichloromethane	PP-ground	82.60	82.15	84.10	82.50	85.17	83.23
Soxhlet	720	90	Dichloromethane	Pellets	53.50	59.40	52.6	60.05	56.17	53.13
Soxhlet	720	90	Dichloromethane	Films	60.60	62.95	65.23	60.53	66.63	62.52
Soxhlet	720	90	Dichloromethane	PP-ground	68.60	69.50	72.67	68.15	69.38	72.80
Ultrasound	90	50	Dichloromethane	Pellets	87.88	89.42	88.04	89.62	88.85	89.41
Ultrasound	90	50	Dichloromethane	Films	92.84	91.76	93.25	91.81	88.47	95.10
Ultrasound	90	50	Dichloromethane	PP-ground	94.68	93.98	95.51	93.68	91.45	96.99
Ultrasound	60	50	Dichloromethane	Pellets	74.10	78.75	77.87	79.25	78.58	72.9
Ultrasound	60	50	Dichloromethane	Films	82.40	84.30	80.37	84.75	79.63	84.51
Ultrasound	60	50	Dichloromethane	PP-ground	87.90	85.50	85.77	91.49	88.53	88.10
Microwave	45	117	Limonene	Pellets	86.07	87.23	84.87	86.08	86.47	87.21
Microwave	45	117	Limonene	Films	90.20	90.80	89.90	90.63	90.00	91.10
Microwave	45	117	Limonene	PP-ground	94.10	89.40	94.27	92.50	94.18	93.85
Microwave	25	117	Limonene	Pellets	74.70	78.80	71.57	78.85	76.50	75.25
Microwave	25	117	Limonene	Films	81.40	82.80	84.33	87.18	85.27	84.10
Microwave	25	117	Limonene	PP-ground	91.55	89.25	91.87	90.88	91.84	92.84
Soxhlet	1440	90	Limonene	Pellets	76.40	72.10	72.53	76.64	71.26	74.43
Soxhlet	1440	90	Limonene	Films	77.80	76.83	78.67	76.08	73.20	79.20
Soxhlet	1440	90	Limonene	PP-ground	79.8	79.93	82.02	80.63	82.57	82.14
Soxhlet	720	90	Limonene	Pellets	52.25	58.45	50.80	59.40	53.78	51.71
Soxhlet	720	90	Limonene	Films	58.80	61.38	62.98	59.64	64.32	61.27
Soxhlet	720	90	Limonene	PP-ground	64.20	67.45	70.50	67.10	68.23	68.09
Ultrasound	90	50	Limonene	Pellets	84.76	83.00	83.65	88.34	88.55	88.20
Ultrasound	90	50	Limonene	Films	83.28	86.7	87.25	85.80	86.63	92.58
Ultrasound	90	50	Limonene	PP-ground	91.08	91.78	91.39	92.33	89.99	96.46
Ultrasound	60	50	Limonene	Pellets	74.10	79.40	74.03	72.03	75.5	76.61
Ultrasound	60	50	Limonene	Films	78.10	77.95	79.67	76.55	80.72	79.74
Ultrasound	60	50	Limonene	PP-ground	83.20	89.05	88.30	84.65	80.27	87.91

2.5. Statistical Analysis

For the present study, a statistical evaluation of the recovery of erucamide in polypropylene samples was conducted. For this purpose, the Minitab software, recognized for its capability of advanced statistical analysis, was used. Given the multifactorial nature of the study, which involved various variables such as extraction techniques (microwave, Soxhlet, and ultrasound), solvents (dichloromethane, cyclohexane, and limonene), and types of polypropylene (ground, pellets, films), a multivariable graphical analysis was performed. This approach allowed for the exploration of interactions and complex relationships among the variables involved in the recovery of erucamide.

Multivariable Graphical Analysis

The multivariable graphical analysis was carried out to visualize and understand the relationship between different extraction techniques, solvents, and forms of polypropylene used in the study. Graphical representation techniques were employed to examine the dependency among variables and explore possible patterns and trends.

3. Results

3.1. Quantification and Recovery of Erucamide by GC-MS

In this study, the presence of erucamide in samples of polypropylene (PP) was determined using a method called the internal standard method. The validity of the GC-MS method was thoroughly tested and confirmed. Both the standard solutions and the samples were analyzed twice to ensure accuracy. The calibration curve, which plots the concentration of erucamide against the instrument response, demonstrated a straight-line relationship within the specified range. The coefficient of determination, a statistical measure of how well the data fit the curve, exceeded 0.999, indicating a highly reliable correlation.

In order to conduct the analysis, we created erucamide solutions with six different concentrations (500, 1000, 1500, 2000, 3000, and 5000 parts per million). These solutions were prepared using an internal standard. The erucamide extracts, obtained from various PP samples such as PP film, PP pellets, and ground PP, were then subjected to analysis using the GC-MS method, following the procedure outlined in Section 2.2.

In the study carried out to evaluate the recovery of erucamide, different extraction techniques (microwave, Soxhlet, and ultrasound), three different solvents (cyclohexane, dichloromethane, and limonene), different forms of polypropylene (films, ground, and pellets) and different extraction times were used. To analyze the results, a variability graph was created to identify the differences in the means and variations in the recovery of erucamide at the combined levels.

Figure 5 shows the relationship between the percentage of erucamide recovery and the variables mentioned above. When analyzing the graph, it was observed that the microwave extraction technique, with a time of 25 min and using ground polypropylene, achieved the highest recovery percentages. However, no significant differences were observed with respect to the solvent used, since the recovery percentages were close to each other. Specifically, dichloromethane obtained a recovery of 92.68%, cyclohexane obtained a recovery of 91.87%, and limonene obtained a recovery of 91.37%. In the case of using an extraction time of 45 min, better results were obtained using ground polypropylene with dichloromethane as a solvent, achieving a recovery percentage of 96.36%. Cyclohexane was in second place with a percentage of 94.57%, and limonene was in last place with 93.05%.

Using ultrasound, better results were obtained in an extraction time of 90 min and likewise in ground pp using dichloromethane as a solvent with a recovery percentage of 94.38%, followed by limonene in ground pp with a percentage of 92.17% and finally cyclohexane in ground pp with a percentage of 87.59%. It can be observed that limonene obtained better results than cyclohexane regardless of the presentation of the polymer. In an extraction time of 60 min, the solvent with the best recovery percentage was dichloromethane in ground pp with 87.88%, followed by limonene with 85.56% and, finally, Cyclohexane with 81.97%.

Finally, Soxhlet extraction obtained lower recovery results compared to microwave and ultrasound. The highest percentage reached using this technique was with a time of 1440 min, using dichloromethane as a solvent and ground pp, with 83.29% recovery. This was followed by limonene in ground pp with 81.18% and lastly by cyclohexane with 78.77%.

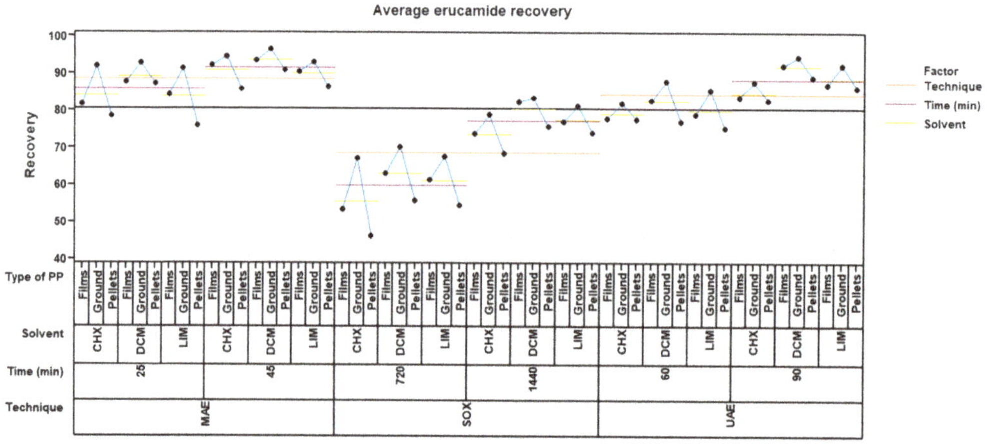

Figure 5. Average recovery of erucamide for microwave (MAE), Soxhlet (SOX), and ultrasound (UAE) with cyclohexane (CHX), dichloromethane (DCM), and limonene (LIM).

This may be because the microwave extraction technique uses microwaves to selectively heat the solvent and sample. This allows for faster and more efficient heat transfer, which speeds up the extraction process. In contrast, the ultrasound and Soxhlet techniques may require more time to reach the right temperature and achieve a complete extraction, as demonstrated in the experimental design, since the Soxhlet extraction needed a time of 1440 min to achieve good recoveries that were well below the recovery percentages obtained by microwave, which only took 45 min. Another reason is that the microwave extraction technique can provide greater agitation and turbulence in the sample, which improves the interaction between the solvent and the analyte. This facilitates erucamide extraction and improves recovery efficiency. And as is known, the microwave extraction technique was able to achieve comparable or better results in a shorter extraction time compared to the ultrasonic and Soxhlet techniques. A shorter extraction time can minimize the degradation or loss of the analyte during the process and improve the recovery, as detailed in Section 2.3, and also allows greater control of the extraction conditions, such as temperature and pressure. This allows the conditions to be optimized to maximize the recovery of the erucamide and minimize any possible interference or degradation of the analyte. Although dichloromethane showed slightly higher recovery percentages, the difference was not significant enough to completely rule out limonene as a solvent option.

In these cases, it is important to weigh the additional benefits of limonene as a green solvent, such as lower toxicity and reduced environmental impact. Furthermore, the choice of solvent depends on other factors, such as current environmental regulations, specific application requirements, and personal or company preferences. If sustainability is valued and the minimization of environmental impact is sought, the choice of limonene as a green solvent may be better aligned with these objectives. In addition to not showing a significant difference compared to dichloromethane, limonene also outperformed cyclohexane in terms of percent erucamide recovery. This is another important consideration when choosing limonene as a solvent.

Cyclohexane, being a toxic solvent, may pose occupational health and safety concerns. In addition, this solvent had a lower contribution to the extraction of erucamide in polypropylene, unlike dichloromethane and limonene. Opting for limonene as a safer and less toxic alternative may be beneficial for both operators and the environment. When selecting solvents, it is essential to consider both extraction efficiency and aspects related to safety and environmental impact. In this case, limonene not only demonstrated erucamide recovery comparable to dichloromethane, but also can avoid the risks associated with the use of cyclohexane.

Nielson performed the extraction and quantification of a series of polyolefin additives including erucamide in low-density polyethylene (LDPE) using a 98:2 methylene chloride/isopropanol mixture as extraction solvent, using ground resins (20 mesh, under liquid nitrogen), for 20 min. The highest percentage of recovery obtained by the author was 91% using a microwave oven. For ultrasonic extraction, he used a 75:25 mixture of methylene chloride/cyclohexane, with which he obtained a 94% recovery of erucamide. The author concluded that the erucamide recoveries are similar and very satisfactory for both extraction techniques.

We can say that both the aforementioned study and the present study achieved quite high recovery percentages using different techniques and solvents. In the Nielson study, recoveries of around 91% were obtained using microwave extraction, and recoveries of around 94% were obtained using ultrasound extraction. In this research, recovery percentages higher than 90% were obtained in most of the conditions evaluated, for microwave, ultrasound, and Soxhlet, using different solvents, extraction times, and polymer forms. It is important to note that the experimental conditions such as the type of polymer used, particle size, solvent mixtures, and extraction times vary between studies. These variations clearly influence the results obtained and make a direct comparison between studies difficult.

3.2. Identification of Erucamide by Mass Spectrometry

The erucamide extraction was performed with the objective of obtaining as much of the original substance as possible without significant contamination. However, if the erucamide has been degraded during the process, the recovery percentages will be lower. The extracted erucamide was analyzed by GC-MS to follow up the original erucamide and observe its transformation into degraded byproducts that may be more difficult to recover or detect during the analysis. The GC-MS analysis mentioned above helps to identify the degraded byproducts and determine whether the erucamide has undergone significant degradation. The data analysis was conducted with the understanding that the compounds being examined are degradation products of erucamide.

To identify and measure these compounds, the fragmentation spectrum of erucamide was used as a reference. This spectrum provided valuable information that aided in the identification and quantification of the degradation products. In this way, it was possible to relate the peaks and features observed in the spectra of the analyzed compounds with the structure and fragments present in the erucamide.

Erucamide is susceptible to oxidative degradation due to its chemical structure and the presence of functional groups. During the extraction process, especially when solvents such as cyclohexane, dichloromethane, and limonene are used, conditions that favor the oxidation of erucamide can occur. The oxidation of erucamide can lead to the formation of degradation products, which could affect the solubility and extraction efficiency of erucamide in the solvents used. These degradation products could have a lower ability to interact with the solvents, resulting in a lower erucamide extraction yield.

In addition, oxidative degradation of erucamide may lead to the formation of compounds with different properties, such as the generation of more polar compounds. These modified compounds could have a lower affinity for the solvents used in the extraction, which would make their separation from the polypropylene resin more difficult and, consequently, could reduce the extraction yield. It is important to note that the oxidative degradation of erucamide can be influenced by several factors, such as temperature, the presence of catalysts, the duration of the extraction process, and the storage conditions of the polypropylene resin. A higher degree of oxidative degradation of erucamide may be indicative of a less efficient extraction process and therefore a lower yield.

To optimize extraction yield, it is important to consider measures to minimize oxidative degradation of erucamide during the extraction process and storage of polypropylene resin. The results of this study indicated that measuring the degree of oxidative degradation of erucamide in the polyolefin resin is indirectly a measure of the erucamide extraction performance imparted to the polyolefin resin. Mass spectrometry techniques were employed to verify the proposed mechanisms responsible for the generation of degradation species. Through the utilization of these techniques, the proposed mechanisms were successfully validated.

Figure 6 shows the mass spectrum obtained, which revealed a characteristic pattern of linear hydrocarbons. To better understand the structure of erucamide and the fragmentation patterns observed, a fragmentation mechanism is recommended in Figure 7. The loss of small molecules, such as hydrogen, results in a decrease in the total amount of erucamide extracted because the more fragmentation occurs and the more molecules are lost, the lower the final amount of erucamide obtained in the extraction process. Table 2 shows the byproducts of erucamide degradation.

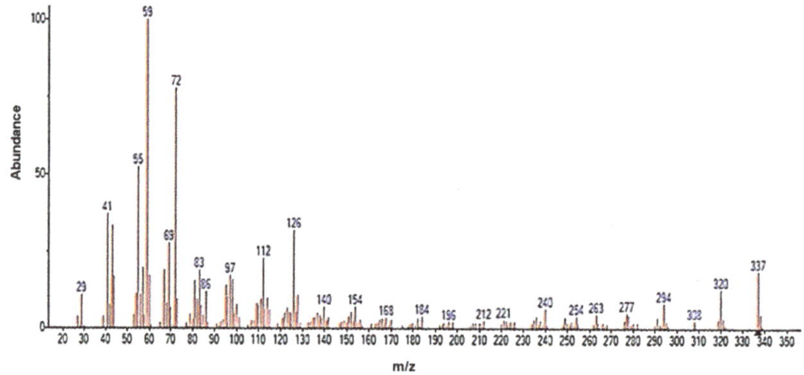

Figure 6. MS spectrum of the recovered erucamide.

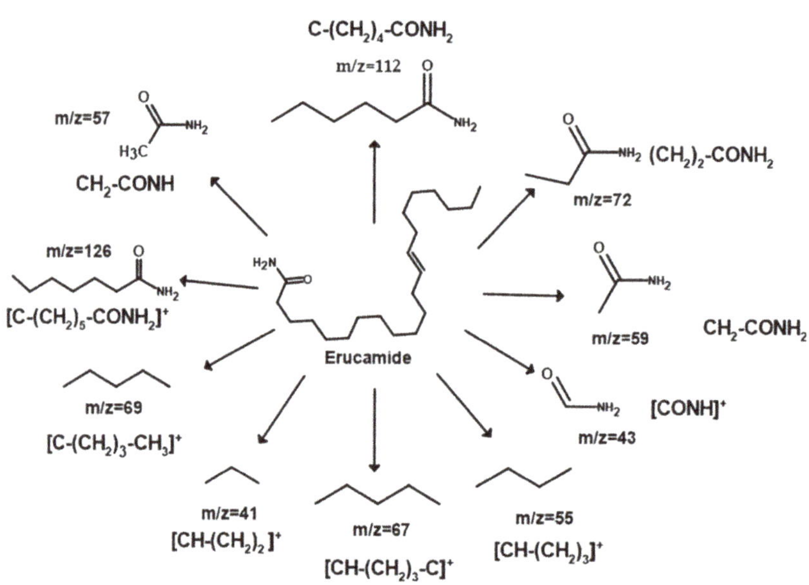

Figure 7. Recovered erucamide fragmentation mechanism.

Table 2. Degradation profile of erucamide in different solvents.

Tr (Min)	Compound	Mass	Formula	Fragments (m/z)	% of Degraded Fragments/Type of Solvent/Extraction Technique					
					Microwave		Soxhlet		Ultrasound	
3.06	12-amino-12-oxo-dodecanoic acid	228.16	$C_{12}H_{23}NO_3$	211.13, 210.15, 184.17, 167.14	DCM	0.15	DCM	0.24	DCM	0.093
					CHX	0.2133	CHX	0.333	CHX	0.15
					LIM	0.128	LIM	0.293	LIM	0.091
4.34	14-amino-14-oxotetradecanoic acid	256.2	$C_{14}H_{27}NO_3$	239.16, 212.2, 238.18 195.17	DCM	0.0693	DCM	0.093	DCM	0.0413
					CHX	0.1173	CHX	0.1466	CHX	0.0703
					LIM	0.124	LIM	0.136	LIM	0.063
6.58	14-oxotetradecanamide	242.2	$C_{14}H_{27}NO_2$	225.18, 207.17, 197.2	DCM	0.126	DCM	0.226	DCM	0.0913
					CHX	0.1363	CHX	0.296	CHX	0.1156
					LIM	0.188	LIM	0.253	LIM	0.118
7.26	15-oxopentadec-13-enamide	254.21	$C_{15}H_{27}NO_2$	239.2, 237.18, 219.17, 201.16	DCM	0.283	DCM	0.443	DCM	0.203
					CHX	0.38	CHX	0.54	CHX	0.286
					LIM	0.29	LIM	0.34	LIM	0.2
7.85	Undecanamide	186.18	$C_{11}H_{23}NO$	169.16, 158.15, 151.08	DCM	0.18	DCM	0.183	DCM	0.08
					CHX	0.266	CHX	0.36	CHX	0.1733
					LIM	0.271	LIM	0.16	LIM	0.124
8.8	Erucamide keto-epoxide	368.3	$C_{22}H_{41}NO_3$	351.29, 352.32, 333.28, 315.27	DCM	0.1896	DCM	0.12	DCM	0.0953
					CHX	0.221	CHX	0.21	CHX	0.159
					LIM	0.316	LIM	0.136	LIM	0.089
10.78	Erucamide with ketone	352.3	$C_{22}H_{41}NO_2$	335.29, 317.28, 307.29, 299.27	DCM	0.203	DCM	0.293	DCM	0.136
					CHX	0.216	CHX	0.4266	CHX	0.19
					LIM	0.22	LIM	0.296	LIM	0.136
11.3	Erucamide with one -OH	354.3	$C_{22}H_{43}NO_2$	337.31, 336.33, 319.29, 309.31, 301.29	DCM	0.26	DCM	0.47	DCM	0.193
					CHX	0.383	CHX	0.546	CHX	0.306
					LIM	0.21	LIM	0.38	LIM	0.506
12	Cis-11-eicosenamide	310.3	$C_{20}H_{39}NO$	293.28, 275.27	DCM	0.129	DCM	0.1613	DCM	0.0736
					CHX	0.1846	CHX	0.2173	CHX	0.1246
					LIM	0.11	LIM	0.206	LIM	0.063
13	Erucamide (13-cis-Docosenamide)	338.3	$C_{22}H_{43}NO$	321.32, 303.31	DCM	93.55	DCM	80.43	DCM	91.819
					CHX	90.804	CHX	73.629	CHX	84.67
					LIM	89.935	LIM	77.343	LIM	88.431

Upon fragmentation, the erucamide leads to the breaking of important bonds in its structure, such as the H-H bond, as observed in the fragmented ions with m/z 41, 55, 59, 72, 112, and 126, which correspond to the fragmentation of the chain near the amide group and the breaking of hydrogen bonds. These bonds are an integral part of the molecule and contain valuable information about its composition and properties. When they are broken, this information is lost and the precise identification of the erucamide and the interpretation of its structure become difficult.

3.2.1. Determination of Thermo-Oxidative Degradation Byproducts of Erucamide

Figures 8–12, which illustrate the mechanisms underlying the formation of erucamide degradation products, are presented below. These mechanisms are characterized by the abundant presence of hydrogen and hydroxyl radicals, the simultaneous occurrence of the generation of these radicals in several different molecules and at different times, and the random nature of the reactions that occur between these radicals and the macro radicals of erucamide. It is important to note that many of these mechanisms present advanced starting species or radicals already formed because the process through which they reach that state, as described in Figure 8, is common to all of them. However, it is necessary to distinguish between the degradation routes that originate in the first part of Figure 8, where the macroradical erucamide and the hydrogen radical are generated; the second part of the route in Figure 8, where, after the union of oxygen and a hydrogen radical to form the peroxide function, the scission of the peroxo (O-O) bond occurs, thus generating the hydroxyl radical and an oxygen radical attached to the carbon chain of the erucamide; and the third part of the mechanism in Figure 8, where the carbon chain of the erucamide is broken, generating a formyl group and an alkenyl (α) free radical or an aldehyde and a radical with the amido (β) group characteristic of erucamide.

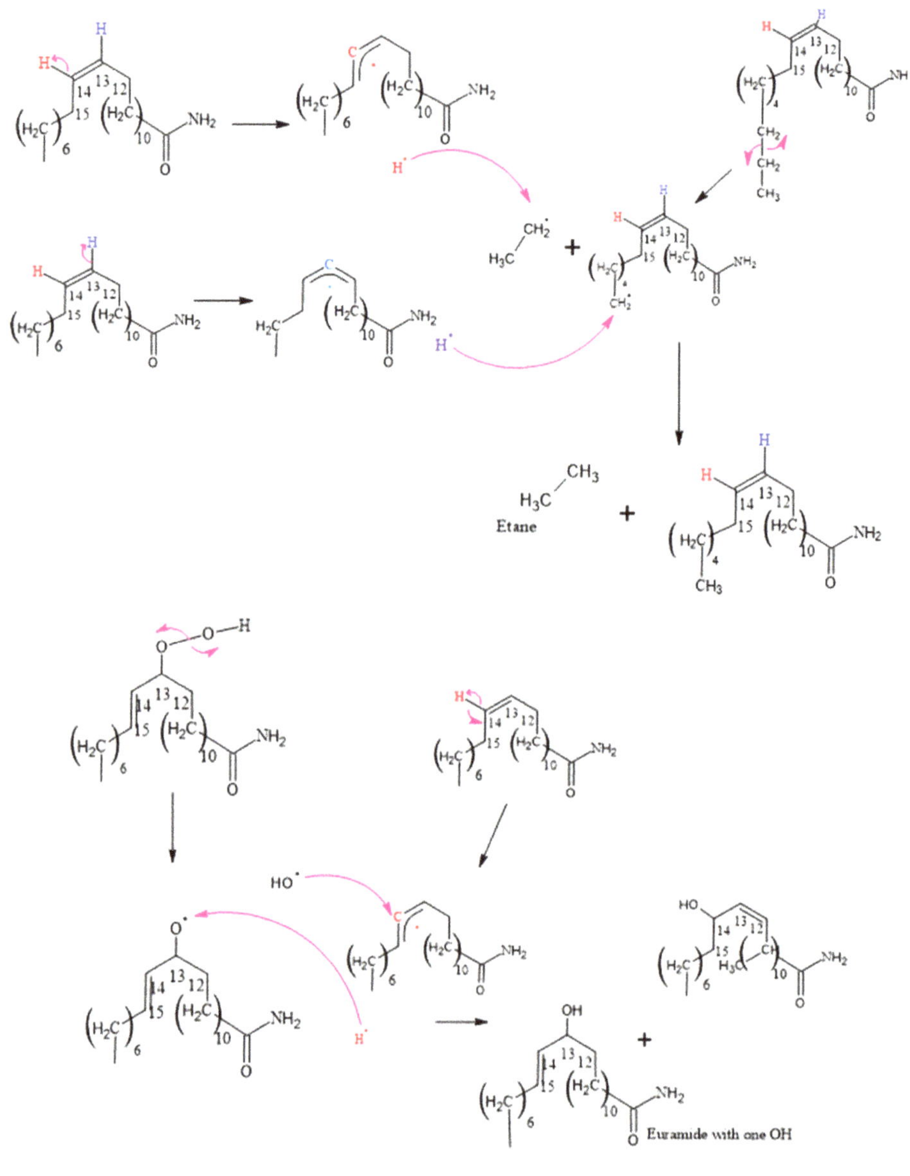

Figure 8. Formation of Cis 11-eicosenamide and erucamide with one OH.

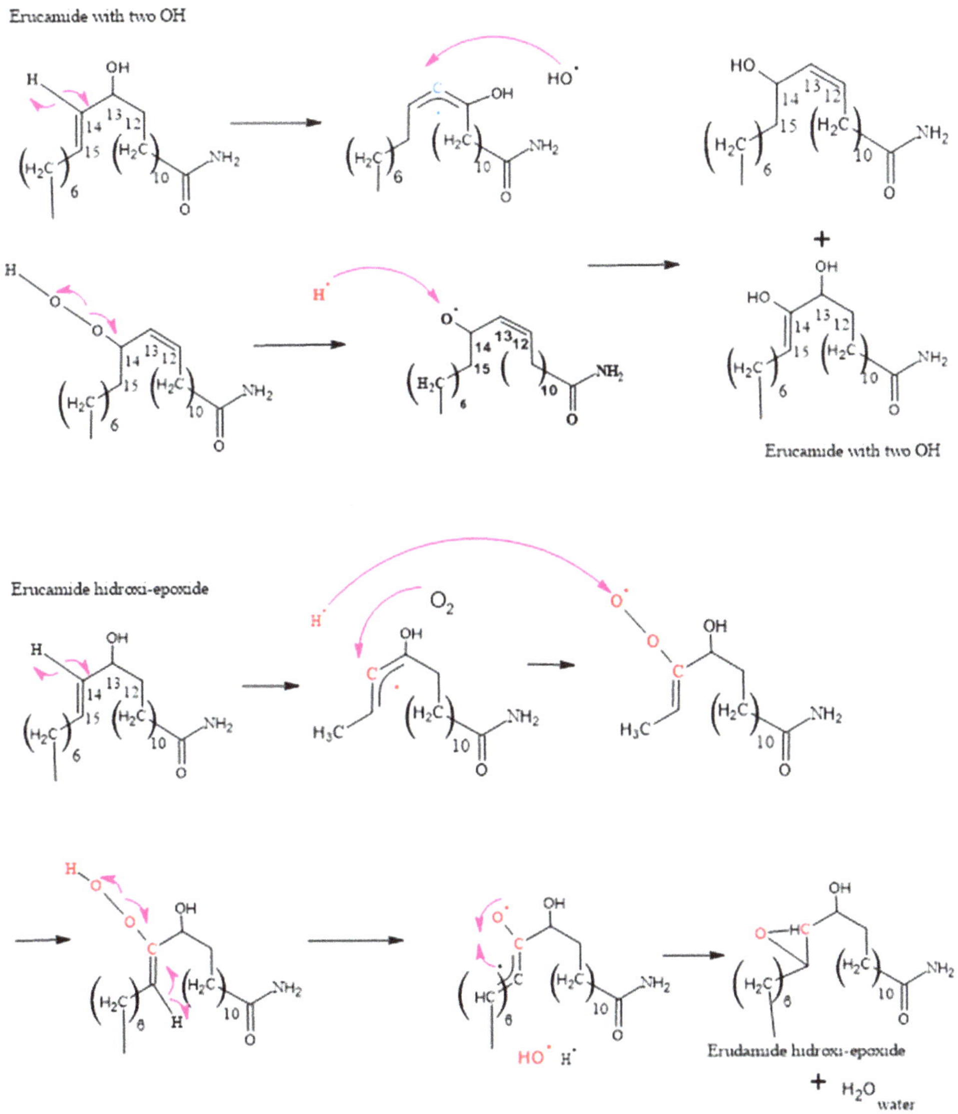

Figure 9. Formation of the degradation of erucamide with two OH groups and erucamide hydroxy-epoxide.

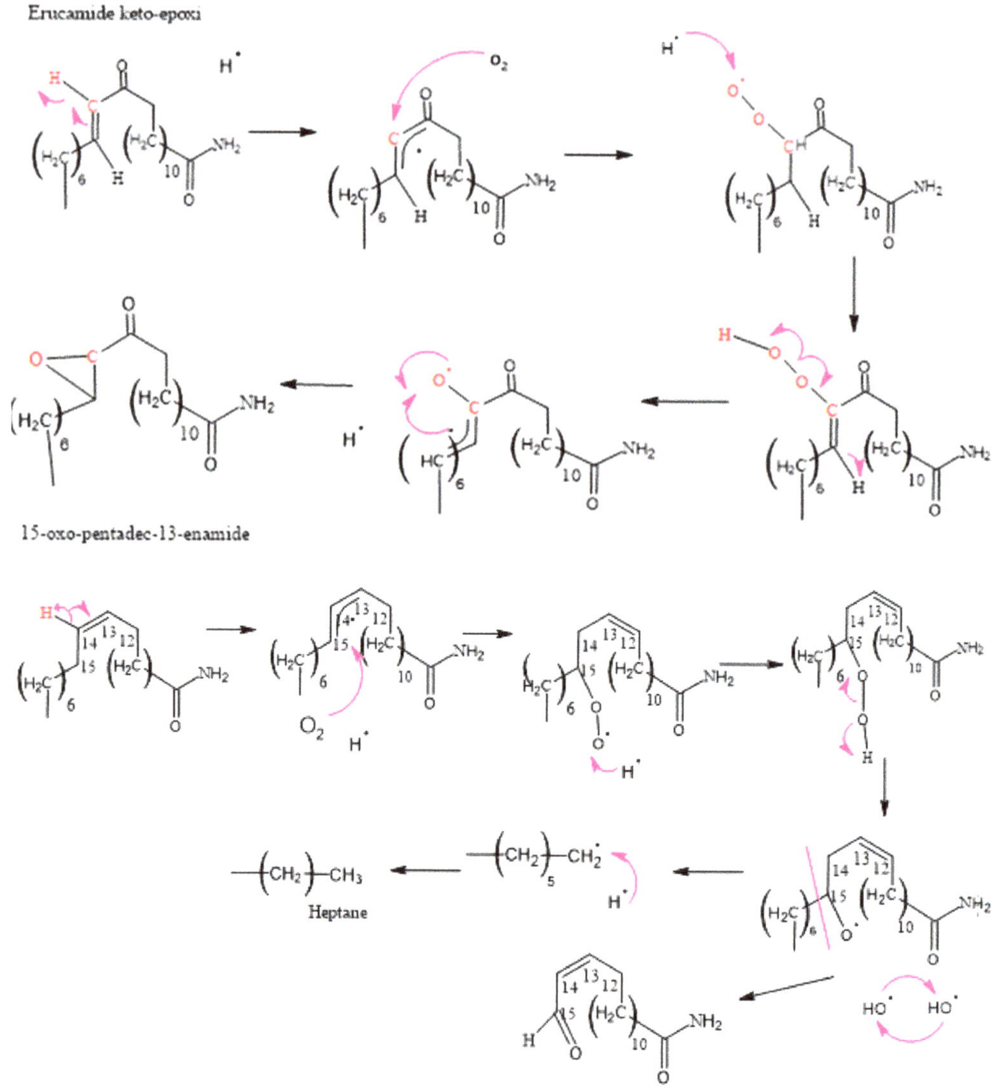

Figure 10. Formation of erucamide keto-epoxide and 15-oxo-pentadec-13-enamide.

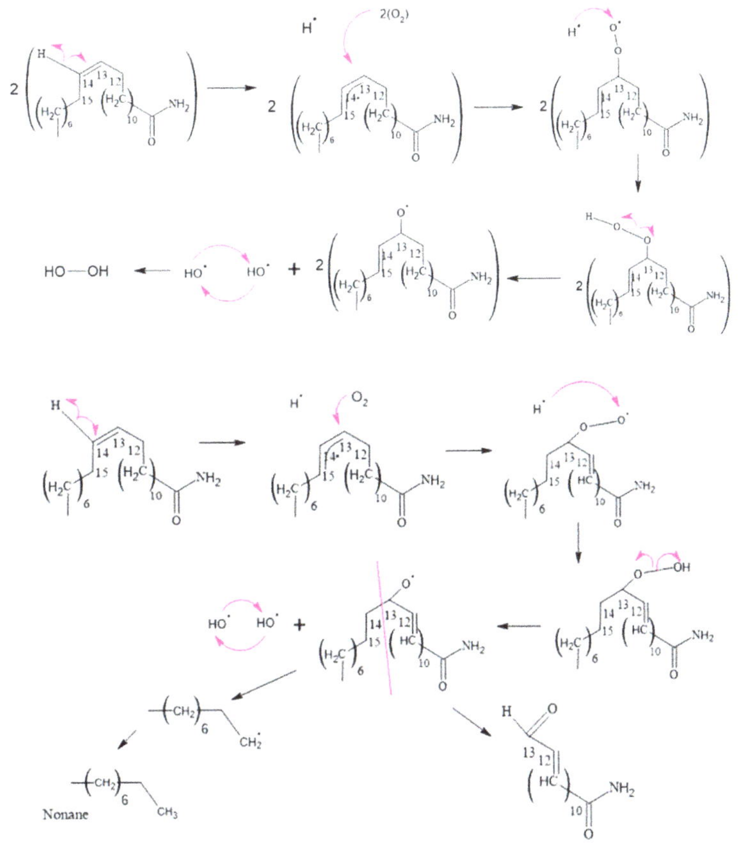

Figure 11. Formation of hydrogen peroxide and 13-oxo-pentadec-11-enamide.

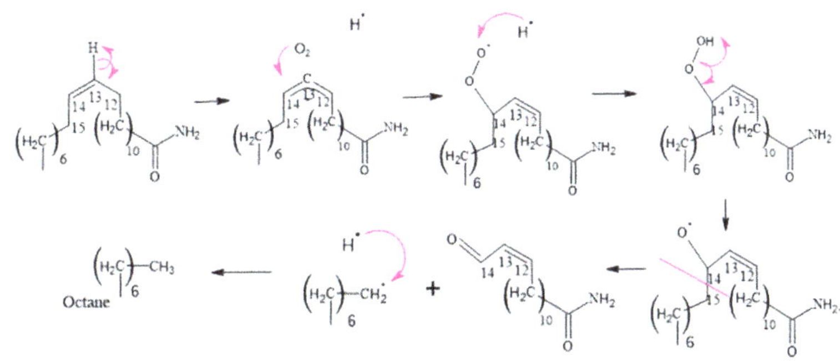

Figure 12. Formation of 14-oxotetradec-12-enamide.

3.2.2. Formation of Erucamide with Two OH Groups and Hydroxy-Epoxide

Figure 9 shows one of the analyses performed on two of the degradation products of erucamide, which allowed us to obtain significant results in terms of understanding the chemical reactions involved. The presence of hydrogen and hydroxyl radicals in abundance suggests that they play a fundamental role in the formation of the degradation products. Moreover, the simultaneity in the generation of these radicals in different molecules and at different times provides an enabling scenario for a series of chain reactions and cascade reactions.

The randomness of the interactions between the hydrogen, hydroxyl, and macro radicals of erucamide introduces a complexity factor into the system, leading to the formation of a variety of degradation products, such as erucamide hydroxy-epoxide. The diversity of the products obtained can be attributed to the different positions at which the radicals bind to the carbon chain of the erucamide, resulting in structural modifications and the formation of new chemical bonds.

3.2.3. Formation of Erucamide Keto-Epoxide and the 15-Oxo-pentadec-13-enamide

It is important to note that, due to the complexity of the mechanisms and chemical reactions involved, the formation of erucamide degradation products does not follow a specific and predictable pattern. Instead, the random nature of the reactions and the interaction between radicals contribute to the diversity of the final products. Therefore, a thorough and detailed experimental approach is required to fully understand the degradation products generated under specific conditions. The compounds shown in Figure 7 display additional product ions where the loss of 18 units (-H_2O) occurs. Additionally, they exhibit characteristic fragments that arise from the presence of amide bonds. Typically, the loss of 18 units is not highly specific because it is commonly observed in compounds that possess functional groups containing oxygen. However, this loss of water molecules is frequent in aliphatic alcohols with a relatively high ratio [55]. Hence, these compounds are generated as a result of the oxidation and degradation process of erucamide. They have a relatively short aliphatic chain and consist of an amide group along with one or more hydroxyl groups.

3.2.4. Hydrogen Peroxide and 13-Oxo-pentadec-11-enamide Formation

The process of lipid oxidation is intricate, and unsaturated fatty acids are especially prone to oxidative degradation. The creation of various degradation products is influenced by multiple factors, including the presence of oxygen, ultraviolet and visible light exposure, heat, and metal catalysts. These factors play a role in the generation of undesirable oxidation products.

In the course of this process, hydrogen peroxides are formed by the extraction of hydrogen from peroxy radicals, as illustrated in Figure 11. Unsaturated fatty acids, with their double bonds, are more susceptible to hydrogen abstraction due to their lower dissociation energy in comparison to saturated aliphatic chains. The instability of hydroperoxides causes them to decompose into alkoxy radicals, which undergo β-scission on both sides of the alkoxy carbon, leading to the formation of aldehydes, ketones, carboxylic acids, alcohols, epoxides, and hydrocarbons.

3.2.5. Formation of 14-Oxotetradec-12-enamide and 15-Amino-15-oxopentadec-2-enoic Acid

When the C14 of erucamide undergoes oxidation, several major products are generated. These include nonanal, 14-oxotetradec-12-enamide, and 13-oxotridecanamide. In addition, 13-oxotridecanamide is converted to 13-amino-13-oxotridecanoic acid, while 14-oxotetradec-12-enamide is converted to 14-amino-14-oxo-tetradec-2-enoic acid.

In Figure 12, an erucamide derivative is shown that features a previously formed formyl group. In this process, the formyl group undergoes a cleavage of the carbon–hydrogen bond within the same group, resulting in the formation of a radical. This radical is in a reactive position and is attacked by a hydroxyl radical from another molecule containing a peroxo bond.

This reaction allows the stabilization of both radicals since a new chemical connection is formed between them, thus generating an acid group. In other words, the formyl group undergoes a cleavage in its carbon–hydrogen bond, which gives rise to a radical that combines with a hydroxyl radical to form a new acid group in the molecule.

3.2.6. Formation of 13-Amino-13-oxotridecanoic Acid and 15-Amino-15-oxopentadeca-enoic Acid

The mechanism illustrated for these compounds observed in Figure 13 is similar to that observed for 15-amino-15-oxo-pentadec-2-enoic acid. However, in this case, it is a chain with 13 and 14 carbons, which is a variant of the two possible carbons where oxygen can attack during the first carbon–hydrogen scission in the double bond of the original erucamide molecule.

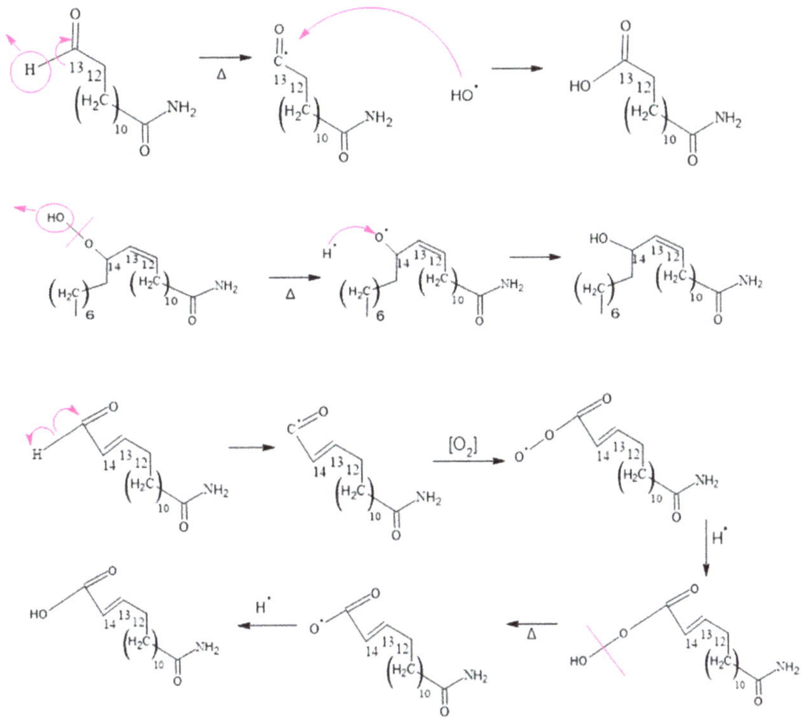

Figure 13. Formation of 13-amino-13-oxotridecanoic and 15-Amino-15-oxopentadeca-enoic Acid.

This mechanism also applies to other similar compounds, such as 12-amino-12-oxododecanoic acid, 14-amino-14-oxotetradecanoic acid and 14-amino-14-oxotetradec-2-enoic acid. In these cases, the difference is in the length of the carbon chain, but the process of carbon–hydrogen scission and formation of amino groups and acids is analogous to that described above. As shown in Figure 14.

3.3. Percentage Analysis of Erucamide Degradation Byproducts

Figure 15 illustrates the variability of different methods, solvents and forms of polypropylene related to erucamide concentrations. The differences in erucamide byproduct concentrations between the Soxhlet and microwave techniques, as well as between the solvents used, can be attributed to the different extraction conditions and the specific properties of the solvents in terms of their ability to extract and retain the degradation byproducts.

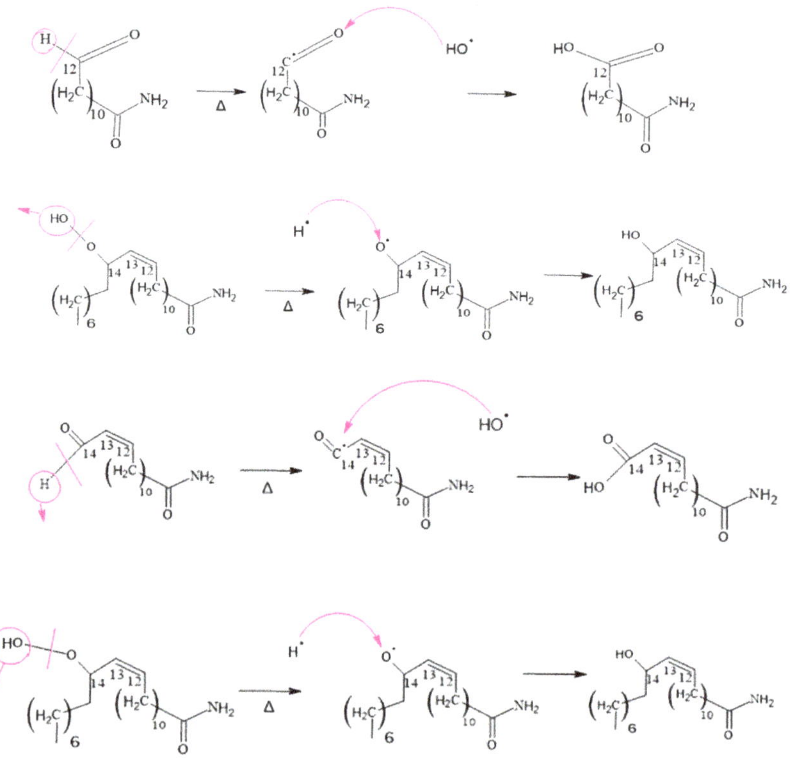

Figure 14. Formation of 12-Amino-12-oxododecaenoic Acid and 14-Amino-14-oxo-tetradec-2-enoic Acid.

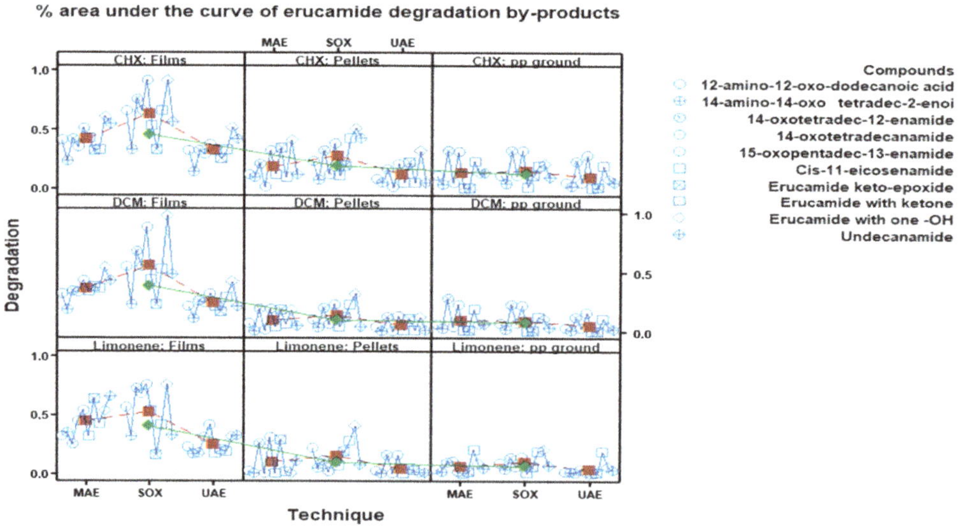

Figure 15. Percent area under the curve of erucamide degradation byproducts without the concentration of erucamide.

In the case of the Soxhlet technique, which involves prolonged and continuous extraction, a higher concentration of erucamide byproducts may have been obtained due to the longer duration and greater contact with the solvent. In addition, cyclohexane and dichloromethane, used as solvents in the Soxhlet technique, may have a greater ability to extract erucamide degradation byproducts compared to limonene.

The difference in the ability of the solvents (cyclohexane, dichloromethane and limonene) to extract erucamide degradation byproducts can be attributed to their chemical and physical properties. Cyclohexane and dichloromethane are organic solvents of medium to low polarity. These solvents are known for their ability to dissolve organic compounds and can efficiently extract erucamide degradation byproducts. This is because these solvents are lipid soluble and have a higher affinity for organic compounds. On the other hand, limonene is a naturally occurring solvent with relatively high polarity. Although limonene can also dissolve organic compounds, its ability to extract erucamide degradation byproducts may be lower compared to cyclohexane and dichloromethane. This is because limonene has a lower solubility in lipids and may have a lower affinity for organic compounds compared to the other solvents mentioned.

On the other hand, the microwave technique generally involves a faster and more efficient extraction due to the use of microwave radiation to heat the sample. Although this technique provided higher erucamide recovery percentages compared to the ultrasound technique, it is possible that the shorter extraction times used in the microwave technique did not allow a complete extraction of the degradation byproducts.

The difference in erucamide degradation rates between polypropylene (PP) in film, milled and pellet form may be related to the accessibility and physical structure of the material. When PP is in film form, it has a larger exposed surface area compared to ground PP and pellets. This means that more surface area is available for solvents and degradation conditions to interact with the erucamide and its byproducts. As a result, it is possible that more degradation of erucamide may occur in the PP in film form, which would be reflected in higher degradation percentages. On the other hand, ground PP and pellets have a more compact physical structure compared to film. This may hinder the access of solvents and degradation conditions to the erucamide and limit the interaction between them. As a result, less degradation of erucamide may occur in the milled PP and pellets, which would be reflected in lower degradation percentages, as shown in Figure 15.

4. Conclusions

The results of the study show that the use of ground polypropylene instead of polypropylene forms in films and pellets can improve erucamide recovery due to higher contact surface area, higher permeability, smaller particle size and higher homogeneity of the material. In addition, the microwave extraction technique was found to be more effective than ultrasound and Soxhlet techniques, as it allowed shorter extraction times and higher recovery efficiency. In microwaves with a time of 25 min and using ground polypropylene, the highest erucamide recovery was achieved, with percentages higher than 96.4%. With a time of 45 min and using ground polypropylene, the recovery reached 96.36%. Using the ultrasound technique, with an extraction time of 90 min and using ground polypropylene, a recovery of 94.38% was obtained. Using the Soxhlet extraction technique, with an extraction time of 1440 min and using ground polypropylene, a recovery of 83.29% was achieved. These results indicate that the microwave extraction technique in combination with ground polypropylene obtained the highest recovery percentages in relatively short extraction times. Although dichloromethane showed a slight advantage in terms of recovery, the use of limonene as a solvent was also viable and offered additional benefits, such as lower toxicity and reduced environmental impact. In contrast, cyclohexane raises occupational health and safety concerns due to its toxicity. Therefore, selecting limonene as a safer and environmentally friendly alternative may be beneficial. In general, it is important to consider both extraction efficiency and safety and environmental impact issues when choosing the appropriate solvent.

Author Contributions: Conceptualization, O.-T.R. and H.F.J.; methodology, P.M.J.; validation, H.F.J., P.M.J. and O.-T.R.; formal analysis, O.-T.R.; investigation, O.-T.R. and H.F.J.; writing—original draft preparation, H.F.J.; writing—review and editing, O.-T.R. and P.M.J.; supervision, H.F.J.; project administration, P.M.J. All authors have read and agreed to the published version of the manuscript.

Funding: This research received no external funding.

Institutional Review Board Statement: Not applicable.

Data Availability Statement: Not applicable.

Conflicts of Interest: The authors declare no conflict of interest.

References

1. Hernández-Fernández, J.; Vivas-Reyes, R.; Toloza, C.A.T. Experimental Study of the Impact of Trace Amounts of Acetylene and Methylacetylene on the Synthesis, Mechanical and Thermal Properties of Polypropylene. *Int. J. Mol. Sci.* **2022**, *23*, 2148. [CrossRef] [PubMed]
2. Hernández-Fernández, J.; Castro-Suarez, J.R.; Toloza, C.A.T. Iron Oxide Powder as Responsible for the Generation of Industrial Polypropylene Waste and as a Co-Catalyst for the Pyrolysis of Non-Additive Resins. *Int. J. Mol. Sci.* **2022**, *23*, 11708. [CrossRef] [PubMed]
3. Hernández-Fernández, J.; Cano, H.; Aldas, M. Impact of Traces of Hydrogen Sulfide on the Efficiency of Ziegler–Natta Catalyst on the Final Properties of Polypropylene. *Polymers* **2022**, *14*, 3910. [CrossRef] [PubMed]
4. Hernández-Fernández, J.; Cano-Cuadro, H.; Puello-Polo, E. Emission of Bisphenol A and Four New Analogs from Industrial Wastewater Treatment Plants in the Production Processes of Polypropylene and Polyethylene Terephthalate in South America. *Sustainability* **2022**, *14*, 10919. [CrossRef]
5. Hernández-Fernández, J.; Puello-Polo, E.; Trilleras, J. Characterization of Microplastics in Total Atmospheric Deposition Sampling from Areas Surrounding Industrial Complexes in Northwestern Colombia. *Sustainability* **2022**, *14*, 13613. [CrossRef]
6. Hernández-Fernández, J.; Guerra, Y.; Puello-Polo, E.; Marquez, E. Effects of Different Concentrations of Arsine on the Synthesis and Final Properties of Polypropylene. *Polymers* **2022**, *14*, 3123. [CrossRef]
7. Hernández Fernández, J.; Cano, H.; Guerra, Y.; Puello Polo, E.; Ríos-Rojas, J.F.; Vivas-Reyes, R.; Oviedo, J. Identification and Quantification of Microplastics in Effluents of Wastewater Treatment Plant by Differential Scanning Calorimetry (DSC). *Sustainability* **2022**, *14*, 4920. [CrossRef]
8. Hernández-Fernández, J. Quantification of arsine and phosphine in industrial atmospheric emissions in Spain and Colombia. Implementation of modified zeolites to reduce the environmental impact of emissions. *Atmos. Pollut. Res.* **2021**, *12*, 167–176. [CrossRef]
9. Gómez-Contreras, P.; Figueroa-Lopez, K.J.; Hernández-Fernández, J.; Cortés Rodríguez, M.; Ortega-Toro, R. Effect of Different Essential Oils on the Properties of Edible Coatings Based on Yam (*Dioscorea rotundata* L.) Starch and Its Application in Strawberry (*Fragaria vesca* L.) Preservation. *Appl. Sci.* **2021**, *11*, 11057. [CrossRef]
10. Fernández, J.H.; Guerra, Y.; Cano, H. Detection of Bisphenol A and Four Analogues in Atmospheric Emissions in Petrochemical Complexes Producing Polypropylene in South America. *Molecules* **2022**, *27*, 4832. [CrossRef]
11. Fernández, J.H.; Rincón, D.; López-Martínez, J. Development and validation of a prototype for the on-line simultaneous analysis of quality caprolactam synthesized on an industrial scale. *MethodsX* **2023**, *10*, 101952. [CrossRef]
12. Pavon, C.; Aldas, M.; Hernández-Fernández, J.; López-Martínez, J. Comparative characterization of gum rosins for their use as sustainable additives in polymeric matrices. *J. Appl. Polym. Sci.* **2022**, *139*, 51734. [CrossRef]
13. Chacon, H.; Cano, H.; Fernández, J.H.; Guerra, Y.; Puello-Polo, E.; Ríos-Rojas, J.F.; Ruiz, Y. Effect of Addition of Polyurea as an Aggregate in Mortars: Analysis of Microstructure and Strength. *Polymers* **2022**, *14*, 1753. [CrossRef] [PubMed]
14. Hernández-Fernández, J.; Lopez-Martinez, J.; Barceló, D. Development and validation of a methodology for quantifying parts-per-billion levels of arsine and phosphine in nitrogen, hydrogen and liquefied petroleum gas using a variable pressure sampler coupled to gas chromatography-mass spectrometry. *J. Chromatogr. A* **2021**, *1637*, 461833. [CrossRef]
15. Llop, C.; Manrique, A.; Navarro, R.; Mijangos, C.; Reinecke, H. Control of the migration behavior of slip agents in polyolefin-based films. *Polym. Eng. Sci.* **2011**, *51*, 1763–1769. [CrossRef]
16. Hernández-Fernández, J.; Puello-Polo, E.; López-Martínez, J. Recovery of (Z)-13-Docosenamide from Industrial Wastewater and Its Application in the Production of Virgin Polypropylene to Improve the Coefficient of Friction in Film Type Applications. *Sustainability* **2023**, *15*, 1247. [CrossRef]
17. Ramírez, M.X.; Hirt, D.E.; Wright, L.L. AFM Characterization of Surface Segregated Erucamide and Behenamide in Linear Low Density Polyethylene Film. *Nano Lett.* **2002**, *2*, 9–12. [CrossRef]
18. Narayana, R.; Mohana, C.; Kumar, A. Analytical characterization of erucamide degradants by mass spectrometry. *Polym. Degrad. Stab.* **2022**, *200*, 109956. [CrossRef]
19. Rawls, A.S.; Hirt, D.E.; Havens, M.R.; Roberts, W.P. Evaluation of surface concentration of erucamide in LLDPE films. *J. Vinyl Addit. Technol.* **2002**, *8*, 130–138. [CrossRef]
20. Nielson, R.C. Extraction and Quantitation of Polyolefin Additives. *J. Liq. Chromatogr.* **1991**, *14*, 503–519. [CrossRef]

21. Markarian, J. Slip and antiblock additives: Surface medication for film and sheet. *Plast. Addit. Compd.* **2007**, *9*, 32–35. [CrossRef]
22. Martinelli, A.B.; Mesquita, F.A. Thermoplastic Resin Composition Comprising a Mixture of Slip Agents and a Mono-Or Co-Extruded, Laminated or Non-Laminated Film. U.S. Patent No. 9,029,447, 30 December 2010.
23. Molnar, N.M. Erucamide. *J. Am. Oil Chem. Soc.* **1974**, *51*, 84–87. [CrossRef]
24. Shuler, C.A.; Janorkar, A.V.; Hirt, D.E. Fate of erucamide in polyolefin films at elevated temperature. *Polym. Eng. Sci.* **2004**, *44*, 2247–2253. [CrossRef]
25. Garrido-López, Á.; Esquiu, V.; Tena, M.T. Determination of oleamide and erucamide in polyethylene films by pressurised fluid extraction and gas chromatography. *J. Chromatogr. A* **2006**, *1124*, 51–56. [CrossRef]
26. Vandenburg, H.; Clifford, A.; Bartle, K.; Garden, L.; Dean, J.; Costley, C. Critical Review: Analytical Extraction of Additives From Polymers. *Analyst* **1997**, *122*, 101R–116R. [CrossRef]
27. Nerín, C.; Salafranca, J.; Cacho, J.; Rubio, C. Separation of polymer and on-line determination of several antioxidants and UV stabilizers by coupling size-exclusion and normal-phase high-performance liquid chromatography columns. *J. Chromatogr. A* **1995**, *690*, 230–236. [CrossRef]
28. Alejandro, J.; Fernandez, H. Process of Extraction, Quantification and Recovery of Additives in Polypropylene with Natural Biodegradable Solvents and Use of the Polypropylene Resulting from the Multiple Extractions. U.S. Patent Application No. 17/630,296, 24 July 2020.
29. Hernández-Fernández, J.; Puello-Polo, E.; Márquez, E. Furan as Impurity in Green Ethylene and Its Effects on the Productivity of Random Ethylene–Propylene Copolymer Synthesis and Its Thermal and Mechanical Properties. *Polymers* **2023**, *15*, 2264. [CrossRef]
30. Hernández-Fernández, J.; Ortega-Toro, R.; Castro-Suarez, J.R. Theoretical–Experimental Study of the Action of Trace Amounts of Formaldehyde, Propionaldehyde, and Butyraldehyde as Inhibitors of the Ziegler–Natta Catalyst and the Synthesis of an Ethylene–Propylene Copolymer. *Polymers* **2023**, *15*, 1098. [CrossRef]
31. Hernández-Fernández, J.; Guerra, Y.; Espinosa, E. Development and Application of a Principal Component Analysis Model to Quantify the Green Ethylene Content in Virgin Impact Copolymer Resins During Their Synthesis on an Industrial Scale. *J. Polym. Environ.* **2022**, *30*, 4800–4808. [CrossRef]
32. Hernández-Fernández, J.; Ortega-Toro, R.; Castro-Suarez, J.R. Quantification of the Synthetic Phenolic Antioxidant Cyanox 1790 in Bottled Water with SPE-HPLC/MS/MS and Determination of the Impact of the Use of Recycled Packaging on Its Generation. *Water* **2023**, *15*, 933. [CrossRef]
33. da Rosa, G.S.; Vanga, S.K.; Gariepy, Y.; Raghavan, V. Comparison of microwave, ultrasonic and conventional techniques for extraction of bioactive compounds from olive leaves (*Olea europaea* L.). *Innov. Food Sci. Emerg. Technol.* **2019**, *58*, 102234. [CrossRef]
34. Bridson, J.H.; Gaugler, E.C.; Smith, D.A.; Northcott, G.L.; Gaw, S. Leaching and extraction of additives from plastic pollution to inform environmental risk: A multidisciplinary review of analytical approaches. *J. Hazard Mater.* **2021**, *414*, 125571. [CrossRef] [PubMed]
35. Salamanca, D.; Dobslaw, D.; Engesser, K.H. Removal of cyclohexane gaseous emissions using a biotrickling filter system. *Chemosphere* **2017**, *176*, 97–107. [CrossRef]
36. Ciclohexano CASRN 110-82-7 | IRIS | EPA de EE. UU., ORD. Available online: https://iris.epa.gov/ChemicalLanding/&substance_nmbr=1005 (accessed on 4 July 2023).
37. Schlosser, P.M.; Bale, A.S.; Gibbons, C.F.; Wilkins, A.; Cooper, G.S. Human Health Effects of Dichloromethane: Key Findings and Scientific Issues. *Environ. Health Perspect.* **2014**, *123*, 114–119. [CrossRef]
38. Sharma, A.; Yu, E.; Morose, G.; Nguyen, D.T.; Chen, W.T. Designing safer solvents to replace methylene chloride for liquid chromatography applications using thin-layer chromatography as a screening tool. *Separations* **2021**, *8*, 172. [CrossRef]
39. Buranarom, A.; Navasumrit, P.; Ngaotepprutaram, T.; Ruchirawat, M. Dichloromethane increases mutagenic DNA damage and transformation ability in cholangiocytes and enhances metastatic potential in cholangiocarcinoma cell lines. *Chem. Biol. Interact.* **2021**, *346*, 109580. [CrossRef]
40. Cayot, N.; Lafarge, C.; Bou-Maroun, E.; Cayot, P. Substitution of carcinogenic solvent dichloromethane for the extraction of volatile compounds in a fat-free model food system. *J. Chromatogr. A* **2016**, *1456*, 77–88. [CrossRef]
41. Cooper, G.S.; Scott, C.S.; Bale, A.S. Insights from Epidemiology into Dichloromethane and Cancer Risk. *Int. J. Environ. Res. Public Health* **2011**, *8*, 3380–3398. [CrossRef]
42. Welton, T. Solvents and sustainable chemistry. *Proc. R. Soc. A Math. Phys. Eng. Sci.* **2015**, *471*, 20150502. [CrossRef]
43. Yasugi, T.; Kawai, T.; Mizunuma, K.; Kishi, R.; Harabuchi, I.; Yuasa, J.; Eguchi, T.; Sugimoto, R.; Seiji, K.; Ikeda, M. Exposure monitoring and health effect studies of workers occupationally exposed to cyclohexane vapor. *Int. Arch. Occup. Environ. Health* **1994**, *65*, 343–350. [CrossRef]
44. Yuasa, J.; Kishi, R.; Eguchi, T.; Harabuchi, I.; Kawai, T.; Ikeda, M.; Sugimoto, R.; Matsumoto, H.; Miyake, H. Investigation on neurotoxicity of occupational exposure to cyclohexane: A neurophysiological study. *Occup. Environ. Med.* **1996**, *53*, 174–179. [CrossRef] [PubMed]
45. Satira, A.; Espro, C.; Paone, E.; Calabrò, P.S.; Pagliaro, M.; Ciriminna, R.; Mauriello, F. The Limonene Biorefinery: From Extractive Technologies to Its Catalytic Upgrading into *p*-Cymene. *Catalysts* **2021**, *11*, 387. [CrossRef]
46. Cui, G.; Yang, X.; Liu, Z.; Wei, M.; Liu, T.; Gu, H.; Yang, L. Potential Use of Limonene as an Alternative Solvent for Extraction of Gutta-Percha from *Eucommia ulmoides*. *ACS Sustain. Chem. Eng.* **2022**, *10*, 11057–11068. [CrossRef]

47. Boukroufa, M.; Boutekedjiret, C.; Chemat, F. Development of a green procedure of citrus fruits waste processing to recover carotenoids. *Resour.-Effic. Technol.* **2017**, *3*, 252–262. [CrossRef]
48. Virot, M.; Tomao, V.; Ginies, C.; Chemat, F. Total Lipid Extraction of Food Using *d*-Limonene as an Alternative to *n*-Hexane. *Chroma* **2008**, *68*, 311–313. [CrossRef]
49. El-Deen, A.K.; Shimizu, K. Application of D-Limonene as a Bio-based Solvent in Low Density-Dispersive Liquid–Liquid Microextraction of Acidic Drugs from Aqueous Samples. *Anal. Sci.* **2019**, *35*, 1385–1391. [CrossRef]
50. Santos, J.; Vladisavljević, G.T.; Holdich, R.G.; Dragosavac, M.M.; Muñoz, J. Controlled production of eco-friendly emulsions using direct and premix membrane emulsification. *Chem. Eng. Res. Des.* **2015**, *98*, 59–69. [CrossRef]
51. Rodríguez-Llorente, D.; Cañada-Barcala, A.; Muñoz, C.; Pascual-Muñoz, G.; Navarro, P.; Santiago, R.; Águeda, I.; Álvarez-Torrellas, S.; García, J.; Larriba, M. Separation of phenols from aqueous streams using terpenoids and hydrophobic eutectic solvents. *Sep. Purif. Technol.* **2020**, *251*, 117379. [CrossRef]
52. Prache, N.; Abreu, S.; Sassiat, P.; Thiébaut, D.; Chaminade, P. Alternative solvents for improving the greenness of normal phase liquid chromatography of lipid classes. *J. Chromatogr. A* **2016**, *1464*, 55–63. [CrossRef]
53. Esteve-Turrillas, F.A.; Armenta, S.; Garrigues, S.; de la Guardia, M. Smart Sorption Materials in Green Analytical Chemistry. In *Green Analytical Chemistry. Green Chemistry and Sustainable Technology*; Springer: Singapore, 2019. [CrossRef]
54. ICSC 0918-D-LIMONENO. Available online: https://www.ilo.org/dyn/icsc/showcard.display?p_card_id=0918&p_version=2&p_lang=es (accessed on 4 July 2023).
55. Holčapek, M.; Jirásko, R.; Lísa, M. Basic rules for the interpretation of atmospheric pressure ionization mass spectra of small molecules. *J. Chromatogr. A* **2010**, *1217*, 3908–3921. [CrossRef]

polymers

MDPI

Article

Synthesis and Properties of Polyamide 6 Random Copolymers Containing an Aromatic Imide Structure

Yingwei Zhang [1,2], Chunhua Wang [1,*], Yong Yi [1], Wenzhi Wang [1] and Jun Yang [1,2,*]

1 National and Local Joint Engineering Research Center of Advanced Packaging Material Research and Development Technology, School of Packaging and Materials Engineering, Hunan University of Technology, Zhuzhou 412007, China; zhangyingwei@stu.hut.edu.cn (Y.Z.); yiyong@stu.hut.edu.cn (Y.Y.); wangwenzhi@hut.edu.cn (W.W.)
2 Zhuzhou Times New Material Technology Co., Ltd., Zhuzhou 412007, China
* Correspondence: wangchunhua@hut.edu.cn (C.W.); yangjun@csrzic.com (J.Y.)

Abstract: In order to adjust the properties of polyamide 6 (PA6) and expand its application, a new strategy of introducing an aromatic imide structure into the PA6 chain through the random copolymerization method is reported. The diimide diacid monomer was first synthesized by the dehydration and cyclization of pyromellitic dianhydride and 6-aminocaproic acid before it reacted with 1,6-hexamethylene diamine to form poly(amide imide) (PAI) salt, and finally synthesized PA6/PAI random copolymers containing an aromatic imide structure by the random copolymerization of ε-caprolactam and PAI salt. The introduction of an aromatic imide structural unit into the PA6 chain could have a great influence on its properties. As the content of PAI increases, the crystallinity (X_c) and melting temperature (T_m) of the PA6/PAI random copolymer gradually decrease, but its glass transition temperature (T_g) increases obviously. When the PAI content is 20 wt%, the copolymer PA6/PAI-20 has the best comprehensive performance and not only has high thermal stabilities but also excellent mechanical properties (high strength, high modulus, and good toughness) and dielectric properties (low dielectric constant and dielectric loss). Moreover, these properties are significantly superior to those of PA6. Such high-performance PA6 random copolymers can provide great promise for the wider applications of PA6 materials.

Keywords: polyamide 6; poly(amide imide); random copolymerization

Citation: Zhang, Y.; Wang, C.; Yi, Y.; Wang, W.; Yang, J. Synthesis and Properties of Polyamide 6 Random Copolymers Containing an Aromatic Imide Structure. *Polymers* **2023**, *15*, 2812. https://doi.org/10.3390/polym15132812

Academic Editors: Leonard Ionut Atanase, Hui Zhao, Bin Wang and Wei Wu

Received: 22 May 2023
Revised: 20 June 2023
Accepted: 21 June 2023
Published: 25 June 2023

1. Introduction

Polyamide 6 (PA6) as an excellent engineering plastic that has been widely investigated and applied in automotive, electrical, packaging, and other industrial fields due to its lightweight, good mechanical properties, excellent chemical resistance, outstanding processability, and low cost [1–5]. With the rising demand for products in diversified markets, the single pure PA6 resin cannot meet all the requirements. In order to adjust its properties and expand its applications, the development of new PA6-modified varieties has become the focus of current research.

Copolymerization is a common modification method in polymer chemistry, and the properties of copolymers can be adjusted according to the need to change the types of comonomers and their relative proportions [6–10]. In order to adjust the properties of PA6, it is necessary to design and synthesize new copolymers by introducing some rigid structures [11]. Aromatic polyimide (PI) is well-known as a high-performance polymer material with excellent mechanical properties and high thermal stability due to its highly rigid chain structure and strong chain interaction [12–16]. Therefore, introducing an aromatic imide structure into the PA6 chain is an effective modification method. Several examples in the literature have reported the linkage of the aliphatic PA6 chain with a rigid aromatic imide structure by block and graft copolymerization [17–21]. For example, PA6-b-PI-b-PA6 copolymers were synthesized using polyimide oligomers that were end-capped with phenyl 4-aminobenzoate to activate the anionic polymerization of ε-caprolactam, and PA6 chains grew from both ends of the polyimide oligomers to form triblock copolymers [19]. All the PA6-b-PI-b-PA6 copolymers had higher tensile moduli and tensile strengths than PA6, but their elongations at break were much lower than PA6. PI-g-PA6 copolymers were synthesized by the polymerization of phenyl 3,5-diaminobenzoate with several diamines and dianhydrides, and the polyimide oligomers containing pendant phenyl ester groups were then used as activators for the anionic polymerization of ε-caprolactam [20]. These block and graft copolymers exhibited good melt processability, and their thermal stability, impact strength, and moisture resistance were dramatically increased by the incorporation of only 5 wt% PI into both the graft and block copolymers. However, the introduction of an aromatic imide structure into the PA6 chain by random copolymerization is rarely reported. Different from other copolymerization strategies, random copolymerization is a one-pot method, where all the copolymerized monomers are added at the same time. These characteristics make random copolymerization particularly suitable for large-scale industrial production due to its simple process and low cost. Moreover, the properties of the random copolymer can be adjusted in a wide range. Therefore, using the random copolymerization method to synthesize the PA6 random copolymer containing an aromatic imide structure is of great significance both in theoretical research and practical applications.

In our previous work, we reported a new method for the synthesis of poly(amide imide)s (PAIs) using diimide diacid (DIDA) monomers containing different rigid aromatic groups between two imide rings to polymerize with 1,10-diaminodecane via a polycondensation process similar to that of PA66 [22]. Inspired by this synthesis, the method of PAIs was combined with our past experience in the synthesis of PA6/66 random copolymers; therefore, in this work, we first synthesized the DIDA monomer by the cyclization reaction of pyromellitic dianhydride (PMDA) and 6-aminocaproic acid and then reacted it with 1,6-hexamethylenediamine (HMDA) to form PAI salt, and finally synthesized PA6/PAI random copolymers containing an aromatic imide structure through the random copolymerization of ε-caprolactam (CL) and PAI salt. Their chemical structures and molecular weights were characterized by FTIR, ^{1}H-NMR, and intrinsic viscosity measurements. Their thermal properties and crystallization behaviors were researched by TGA, DSC, WAXD, and DMTA. In addition, their rheological behaviors and mechanical and dielectric properties were also systematically investigated. Our results demonstrate how the rigid aromatic imide structure influences the properties of PA6/PAI random copolymers by comparing it with pure PA6.

2. Materials and Methods

2.1. Materials

Pyromellitic dianhydride (PMDA, 99%), 6-aminocaproic acid (98%), 1,6-hexa- methylenediamine (HMDA, 99%), N,N-dimethylformamide (DMF, 99%) and sodium acetate (NaAc, 99%) were purchased from Shanghai Aladdin Biochemical Technology Co., Ltd., Shanghai, China. Acetic anhydride (Ac2O, 99%) was purchased from Sinopharm Chemical Reagent Co., Ltd., Beijing, China. ε-Caprolactam (CL, 99%) was produced in Yueyang Chemical Fiber Co., Ltd., Yueyang, China. The materials were used directly with no further purification.

2.2. Synthesis of DIDA Monomer

PMDA (218.0 g, 1.0 mol), 6-aminocaproic acid (275.1 g, 2.1 mol), and 1500 mL DMF were added into a dried reaction kettle and reacted at room temperature for 1 h. Then, NaAc (8.2 g, approximately equal to 0.1 mol) and Ac2O (10.2 g, approximately equal to 0.1 mol) were added to this reaction kettle. Then, the material temperature in the reaction kettle was raised to 140 °C and maintained for 6 h. Finally, the crude product was filtered and then vacuum-dried under 80 °C for 12 h. The white powder DIDA was synthesized. Yield: 386.3 g, 87%. The synthetic process of the DIDA monomer is displayed in Scheme 1.

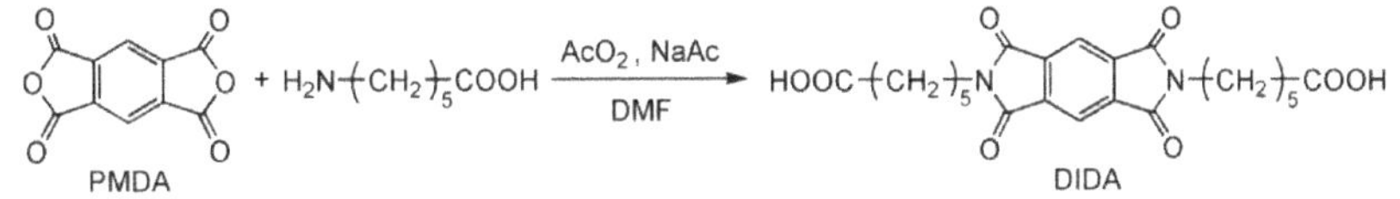

Scheme 1. Synthetic route of DIDA monomer.

2.3. Synthesis of PAI Salt

In total, 500 g of DIDA and 1000 mL of deionized water were added into a salt-forming kettle. HMDA was slowly added into this salt-forming kettle under stirring at 60 °C. When the pH value of the mixture reached 7.5–7.8, we stopped adding HMDA. The mixture stood for 4 h to make the PAI salt precipitate out before PAI salt was obtained by suction filtration and vacuum drying.

2.4. Synthesis of PA6/PAI Random Copolymers

PA6/PAI random copolymers containing an aromatic imide structure were polymerized through the process shown in Scheme 2. The following is a typical process of polymerization. A total of 1600 g of CL and 400 g of PAI salt were mixed with a given mass of deionized water. Then, the mixture was added into an autoclave with a volume of 5 L. The autoclave was filled with high-purity N_2 until the pressure reached 0.3 MPa, and was then placed in a pumping vacuum. This process was repeated three times. The system reacted at 220 °C for 2 h; meanwhile, water vapor was released slowly to control the pressure inside the autoclave so that it did not exceed 2.5 Mpa. Then, the pressure was reduced from 2.5 Mpa to atmospheric pressure by releasing the water vapor, which was then further reduced to −0.05 Mpa by dehydrating in the vacuum. The reaction continued for a further 2 h at 250 °C and −0.05 Mpa. Finally, the polymerzation was finished, and the crude polymer containing the oligomers was extracted in boiling water for 8 h. After that, the saturated water-absorbing polymer was placed in a vacuum drying oven at 80 °C for 12 h to obtain:

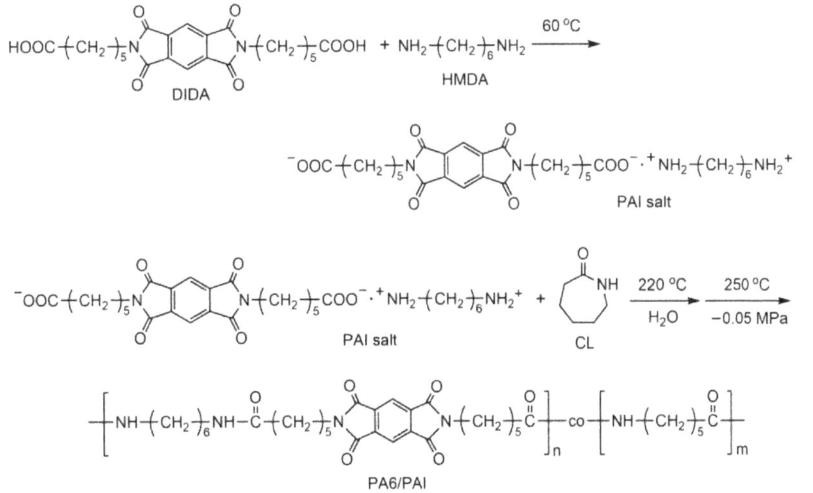

Scheme 2. Synthetic route of PAI salt and PA6/PAI random copolymers.

PA6/PAI-20 random copolymer. Other PA6/PAI random copolymers with a different PAI content, as described in Table 1, were synthesized by the same preparation process.

Table 1. The synthesis of PA6/PAI random copolymers with different PAI content.

Sample	CL (g)	PAI Salt (g)	$[\eta]$ [a] (dL/g)	$\overline{M_\eta}$ [a] ($\times 10^4$)
PA6	2000	0	1.02	2.86
PA6/PAI-5	1900	100	0.95	2.62
PA6/PAI-10	1800	200	0.82	2.19
PA6/PAI-20	1600	400	0.86	2.32
PA6/PAI-30	1400	600	0.91	2.49

[a] $[\eta]$ was the intrinsic viscosity, and $\overline{M_\eta}$ was the viscosity-average molecular weights.

2.5. Characterization

Fourier transform infrared (FTIR) spectra were measured with a Nicolet IS 10 Fourier transform infrared spectrometer with attenuated total reflection (ATR) technology. ^{1}H-NMR spectra were acquired on a Bruker ARX400 spectrometer at room temperature using deuterated dimethyl sulfoxide (DMSO-d_6) or deuterated trifluoroacetic acid (CF_3CO_2D) as the solvent for the DIDA monomer and polymers, respectively. Tetramethylsilane (TMS) was used as the internal reference.

Intrinsic viscosity measurements were tested with a Ubbelohde viscometer at 25 °C using 96% H_2SO_4 as a solvent, and the concentration of the polymer solution was 0.5 g/dL. The intrinsic viscosity $[\eta]$ of the sample was calculated using the following Equation (1):

$$[\eta] = \frac{\sqrt{2(\eta_{sp} - In\eta_r)}}{C} \quad (1)$$

where $\eta_r = t/t_0$, $\eta_{sp} = t/t_0 - 1$. t was the efflux time of the solution, t_0 was the efflux time of the solvent, η_r was the relative viscosity, η_{sp} was the specific viscosity, and C was the concentration of the polymer solution.

Thermogravimetric analysis (TGA) was examined on a Mettler Toledo TGA/DSC1 1100SF instrument. The sample was heated from 25 °C to 700 °C at a rate of 10 °C/min in a nitrogen atmosphere. Differential scanning calorimeter (DSC) thermograms were detected using a Netzsch DSC 204 F1 at a heating and cooling rate of 10 °C/min from 25 °C to 300 °C in a nitrogen atmosphere. The crystallinity (X_c) of the sample was calculated by the following Equation (2):

$$X_c = \frac{\Delta H_m}{\Delta H_m^0} \times 100\% \quad (2)$$

where ΔH_m was the melting enthalpy of the sample obtained by integrating the melting peak of the DSC curve during the second heating process, and ΔH_m^0 was the melting enthalpy of the 100% crystallized PA6 (230.1 J/g) [23].

The wide-angle X-ray diffraction (WAXD) patterns were recorded on a Bruker D8 Discover diffractometer in the 2θ range of 3–40°.

The glass transition temperature (Tg) was measured by a Q850 dynamic mechanical thermal analyzer (DMTA) at a heating rate of 5 °C/min from −70 °C to 100 °C.

Rheological properties were investigated with a Thermo Fisher Mars III rotational rheometer. The specimens with a thickness of 1 mm and a diameter of 25 mm were obtained by injection molding. Complex viscosities (η^*) were tested between parallel plates at 240 °C, and angular frequencies ranged from 0.05 to 500 rad/s.

Tensile strength, Young's modulus, and fracture elongation were detected with the method of GB/T 1040.2-2006, which was equivalent to ISO 527-2:1993. The bending strength and bending modulus were measured according to GB/T 9341-2008, which was equivalent to ISO 178:2001. The impact strength was measured according to GB/T 1043.1-2008, which is equivalent to ISO 179-1:2000. The temperature and relative humidity for the test was

23 ± 2 °C and 50 ± 10%, respectively. Each test was repeated five times, and the average value was calculated together with the standard deviation.

The dielectric constant and dielectric loss were tested using a Subo Electric QS37 high voltage bridge measurement system according to GB/T 1049-2006. The test voltage was 1000 V, the frequency was 50 Hz, and the temperature was 23 ± 2 °C.

3. Results and Discussion

3.1. Structure Characterization of DIDA Monomer

As shown in Scheme 1, the DIDA monomer was synthesized by the dehydration cyclization of PMDA and 6-aminocaproic acid, and its structure was characterized by FTIR and ^{1}H-NMR. Figure 1a shows the FTIR spectrum of the DIDA monomer. It can be observed that the characteristic absorption peaks appeared at 1712 cm^{-1} (C=O, imide ring) and 1398 cm^{-1} (C–N stretching, imide ring), which indicated the successful formation of the imide ring [24,25]. The ^{1}H-NMR spectrum of the DIDA monomer in DMSO-d6 is shown in Figure 1b. The characteristic resonance peaks at 3.48–3.68 ppm were the proton peaks of the methylene connected with the imide group (denoted f), which suggested that the dehydration cyclization reaction was carried out, as described in Scheme 1.

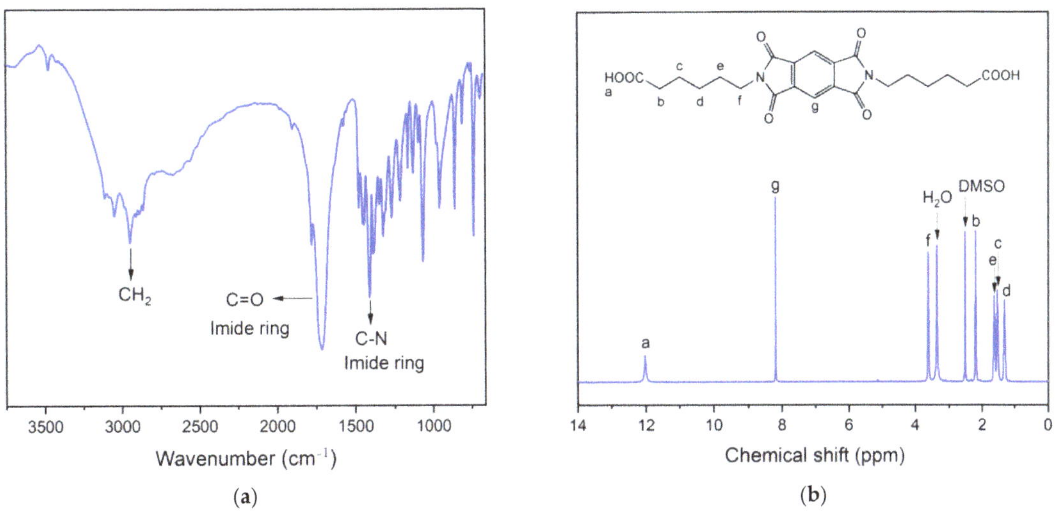

Figure 1. FTIR spectrum (**a**) and ^{1}H-NMR spectrum (**b**) of DIDA monomer.

3.2. Structure Characterization of PA6/PAI Random Copolymers

PA6/PAI random copolymers with a different PAI content were obtained by the random copolymerization of CL and PAI salt, and their polymerization process was similar to that of PA6/66 random copolymers. Deionized water was used as the reaction medium because it is more environmentally friendly and cheaper than organic solvents. The system reacted at 220 °C for 2 h, where CL underwent a hydrolytic ring-opening reaction to form 6-aminocaproic acid, and 6-aminocaproic acid and PAI salt was converted to the oligomer at 220 °C and 2.5 MPa. The degree of polymerization of the oligomer gradually increased with the water vapor. In order to further improve the degree of polymerization, the reaction was continued for a further 2 h at 250 °C and −0.05 MPa. After the reaction ended, PA6/PAI random copolymers were obtained.

The chemical structures of PA6/PAI random copolymers were verified by FTIR and ^{1}H-NMR. Figure 2a shows the FTIR spectra of PA6/PAI random copolymers. The spectral features at 1640 cm^{-1} (C=O, amide) and 1542 cm^{-1} (N-H, C-N, amide) directly verified the formation of an amide bond. Furthermore, the characteristic absorption peak of the imide ring of PA6/PAI random copolymers could still be observed at 1712 cm^{-1}, and the intensity of this peak increased with the increasing PAI content.

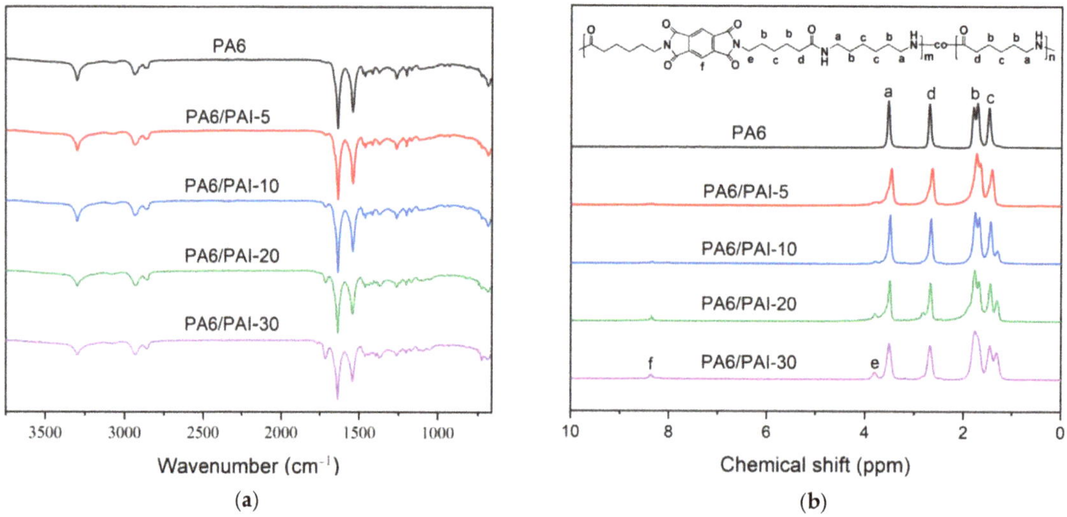

Figure 2. FTIR spectra (**a**) and ^{1}H-NMR spectra (**b**) of PA6/PAI random copolymers.

Figure 2b shows the 1H-NMR spectra of PA6/PAI random copolymers in CF3CO2D. The methylene proton peaks connected with the amide group (denoted a) and connected with the imide group (denoted e), respectively, appeared at 3.36–3.68 ppm and 3.70–3.90 ppm, which demonstrated the existence of amide bonds and imide rings in PA6/PAI random-copolymers. Furthermore, the chemical shifts of other protons in ^{1}H-NMR spectra were also consistent with the designed PA6/PAI random copolymers, and the strength of the methylene proton peak (denoted e) when connected with imide group and benzene proton peak (denoted f) increased with the increase in the PAI content. Therefore, it could be inferred by the FTIR and ^{1}H-NMR spectra that the chemical structures of PA6/PAI random copolymers were as expected.

The viscosity-average molecular weights ($\overline{M_\eta}$) of PA6/PAI random copolymers are usually calculated by intrinsic viscosities $[\eta]$ according to the Mark–Houwink equation, $[\eta] = K\overline{M_\eta}^{\alpha}$, where K and α are constants that only depend on the properties of the polymer solution. However, PA6/PAI random copolymers are new polymers, and their values of K and α are not available in the literature. Therefore, we could only estimate the $\overline{M_\eta}$ of the PA6/PAI random copolymers based on the K and α of PA6, and its values were 2.26×10^{-4} and 0.82, respectively [26–28]. As shown in Table 1, the measured $[\eta]$ values of the PA6/PAI random copolymers ranged from 0.82 to 0.95 dL/g, and the calculated $\overline{M_\eta}$ values were all greater than 21,000, which indicated that the PA6/PAI random copolymers with high molecular weights could be successfully synthesized by the random copolymerization method.

Although the molecular weights of PA6/PAI random copolymers were slightly different, this had little effect on the performance of the materials. Therefore, in the following work, we mainly focused on the influence of the PAI content and on the properties of PA6/PAI random copolymers.

3.3. Thermal Properties and Crystallization Behaviors

The thermal stabilities of PA6/PAI random copolymers were investigated by TGA. Figure 3 demonstrates the TGA graphs of PA6/PAI random copolymers in the N_2 atmosphere, and the initial thermal decomposition temperatures (T_d) at 5% weight loss and char yields at 700 °C are listed in Table 2. Compared with pure PA6, PA6/PAI random copolymers exhibited better thermal stabilities, and the T_d value and char yield of the PA6/PAI random copolymer increased obviously with the increase in the PAI content. The results indicate that the introduction of an aromatic imide structure into the PA6 chain improved its thermal stability.

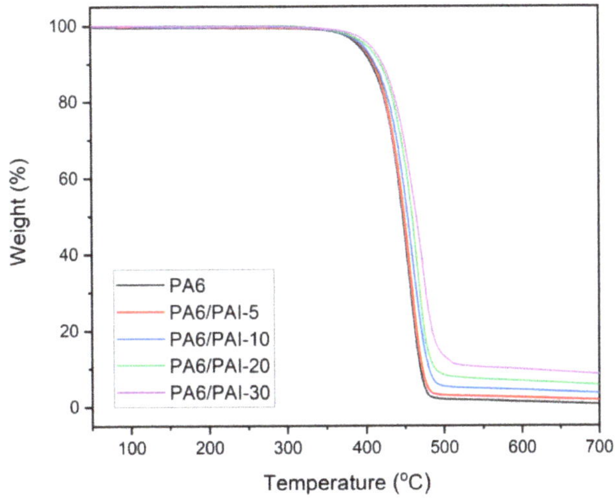

Figure 3. TGA curves of PA6/PAI random copolymers.

Table 2. Thermal properties of PA6/PAI random copolymers.

Sample	T_d [a] (°C)	Char Yield [a] (%)	T_c [b] (°C)	T_m [c] (°C)	ΔH_m [c] (J/g)	X_c [d] (%)	T_g [e] (°C)
PA6	392.1	0.6	182.2	221.8	58.0	25.2	60.5
PA6/PAI-5	393.8	1.7	176.4	217.7	55.7	24.2	63.8
PA6/PAI-10	395.7	3.5	168.2	211.1	54.1	23.5	67.4
PA6/PAI-20	400.2	5.6	158.3	202.1	49.8	21.6	73.1
PA6/PAI-30	405.4	8.4	145.9	190.5	36.5	15.9	77.2

[a] The initial thermal decomposition temperatures T_d at 5% weight loss and char yields at 700 °C were measured by TGA at a rate of 10 °C/min under N_2. [b] The crystallization temperature T_c was obtained by DSC during the cooling process. [c] The melting point T_m and the melting enthalpy ΔH_m were obtained by DSC during the second heating process. [d] $X_c = \Delta H_m / \Delta H_m^0$, $\Delta H_m^0 = 230.1$ J/g. [e] The glass transition temperature (T_g) was obtained by DMTA at a heating rate of 5 °C/min from −70 °C to 100 °C.

The DSC curves of PA6/PAI random copolymers during the first cooling and subsequent heating process are shown in Figure 4, and the characteristic parameters are listed in Table 2. The content of PAI had a great influence on the melting and crystallization behaviors of PA6/PAI random copolymers. As shown in Figure 4, the crystallization temperature (T_c) and melting point (T_m) of the PA6/PAI random copolymer gradually decreased with the increase in the PAI content. The T_c and T_m of PA6 were 182.2 °C and 221.8 °C, while the T_c and T_m of PA6/PAI-30 were 145.9 °C and 190.5 °C. These results suggest that the more aromatic imide groups that were introduced into the PA6 chain, the more irregular the structure of the copolymer was, the ordered arrangement of the chain segments was more difficult, and fewer crystals were formed. Moreover, with the increase in the PAI content, the melting peak of the PA6/PAI random copolymer became ever lower and wider.

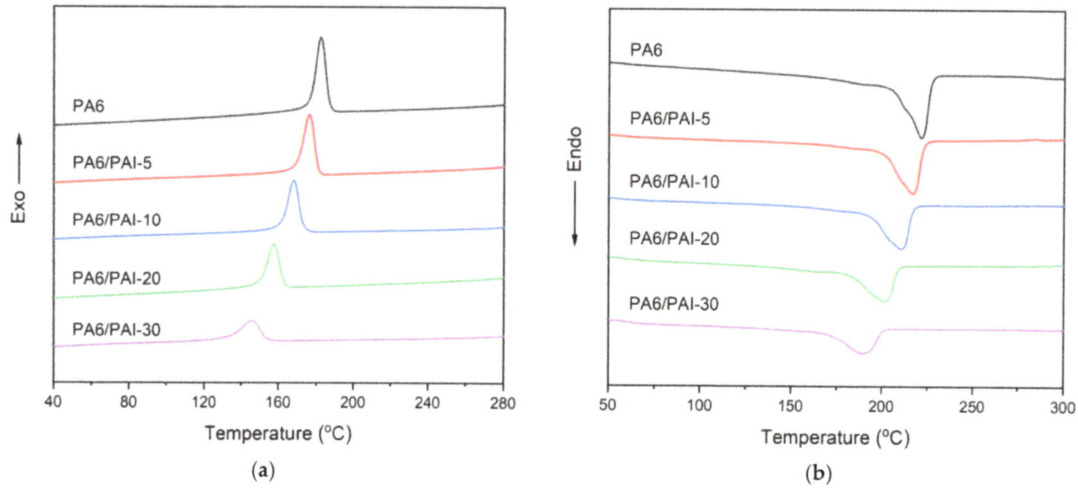

Figure 4. DSC curves of PA6/PAI random copolymers during the first cooling (**a**) and the subsequent heating (**b**) at a rate of 10 °C min^{-1}.

The melting enthalpy (ΔH_m) obtained by integrating the melting peak of the DSC curve during the second heating process is listed in Table 2. With the aid of the melting enthalpy of 100% crystallized PA6 (230.1 J/g), the crystallinity (X_c) was calculated by Equation (2). As shown in Table 2, the X_c of the PA6/PAI random copolymer was obviously lower than that of pure PA6, and the X_c of PA6/PAI random copolymer decreased with the increasing content of PAI, especially in the region with a high PAI content. The variation trend of X_c was the same as that of T_c and T_m. Therefore, the introduction of an aromatic imide structure into the PA6 chain limited the flexibility and mobility of the chain segment and reduced the crystallization ability of the PA6/PAI random copolymer.

The crystal structure of PA6/PAI random copolymers was researched by WAXD, and these curves are demonstrated in Figure 5. Pure PA6 had two diffraction peaks at 2θ = 20.08° and 23.72°, which showed that the crystal structure belonged to α-form. Compared with pure PA6, the positions of these two diffraction peaks of PA6/PAI random copolymer were:

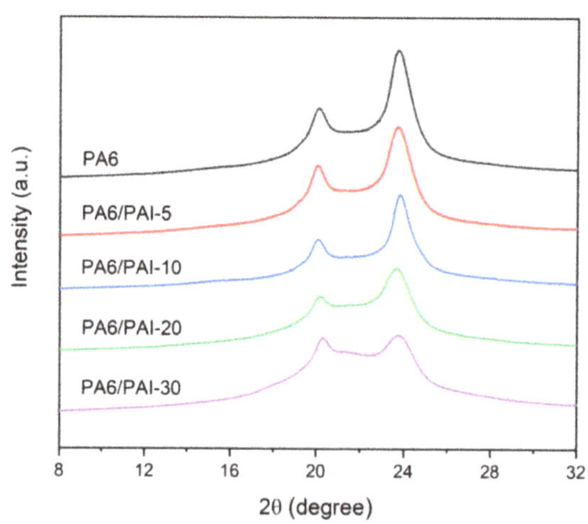

Figure 5. WAXD patterns of PA6/PAI random copolymers.

It was almost crystal in form but led to a decrease in crystallinity. The introduction was unchanged, but their intensities decreased significantly with the increasing PAI content. The results indicate that the introduction of the aromatic imide structure did change the aromatic imide structure breaks of the original regular chain structure of PA6, making it difficult to arrange the chain segments into the lattice regularly. The results of WAXD were consistent with those of DSC.

The glass transition temperature (T_g) of PA6/PAI random copolymers was measured by DMTA. Figure 6 shows the curves of tan δ relating to the temperature for PA6/PAI random copolymers, and the T_g values are listed in Table 2. Note that the T_g values of PA6/PAI random copolymers were obviously higher than that of PA6, indicating the improved rigidity of the chain segment due to the introduction of an aromatic imide structure. The higher the PAI content, the higher the T_g value of PA6/PAI random copolymers. The change in the chain segment rigidity had a great influence on the rheological properties, mechanical properties, and dielectric properties of PA6/PAI random copolymers, and the latter two properties were also affected by the crystallinity.

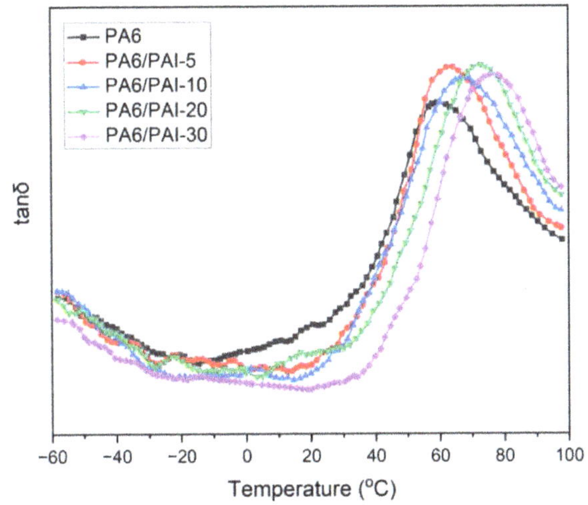

Figure 6. The curves of tan δ relating to the temperature for PA6/PAI random copolymers.

3.4. Rheological Properties

Rheological properties are important for polymer processing and are usually researched by a rotational rheometer. Figure 7 shows the complex viscosity (η^*) of PA6/PAI random copolymers with different PAI content. As the frequency increased, all the samples exhibited the shear thinning behavior of the pseudoplastic fluid, which could be attributed to the molecular chains' orientation along the flow direction, which was untangled at a high shear frequency [29,30]. Compared with PA6, the η^* of PA6/PAI random copolymer increased significantly due to the introduction of the rigid aromatic imide structure. Moreover, the content of PAI had a great effect on the η^* of the PA6/PAI random copolymer. With the increase in the PAI content, the η^* of the PA6/PAI random copolymer increased. The results indicate that the more rigid the aromatic imide groups in the copolymer, the more difficult the movement of the copolymer chain was [31]. Despite the increased viscosities of PA6/PAI random copolymers, they could be processed at lower temperatures due to their low melting points.

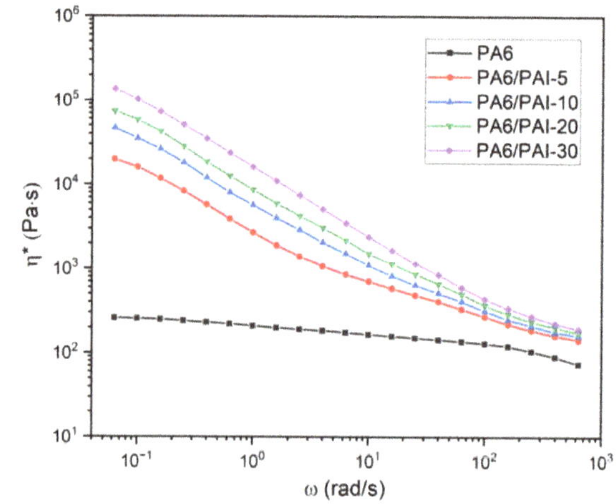

Figure 7. Complex viscosities (η^*) of PA6/PAI random copolymers.

3.5. Mechanical Properties

The mechanical properties of PA6/PAI random copolymers with a different proportion of PAI are demonstrated in Figures 8 and 9. When the content of PAI was 20 wt% or less, the tensile strength, Young's modulus, bending strength, and bending modulus of the PA6/PAI random copolymer obviously increased with the increasing PAI content, and these mechanical properties were significantly higher than those of PA6, which indicated that the introduction of an aromatic imide structural unit improved the rigidity of the chain segment. When the content of PAI reached 30 wt%, these mechanical properties appeared to decline because excessive aromatic imide groups seriously destroyed the crystallization of the PA6 chain segment. Meanwhile, the impact strength and fracture elongation of the PA6/PAI random copolymer also increased first and then decreased and reached the maximum when the PAI content was 20 wt%. A few aromatic imide groups acted as a plasticizer in PA6, improving the plasticity of the copolymer. However, excess aromatic imide groups can also greatly reduce the slip ability between molecular chains, and the toughness of PA6/PAI-30 can deteriorate significantly. Therefore, PA6/PAI-20 had the best mechanical properties.

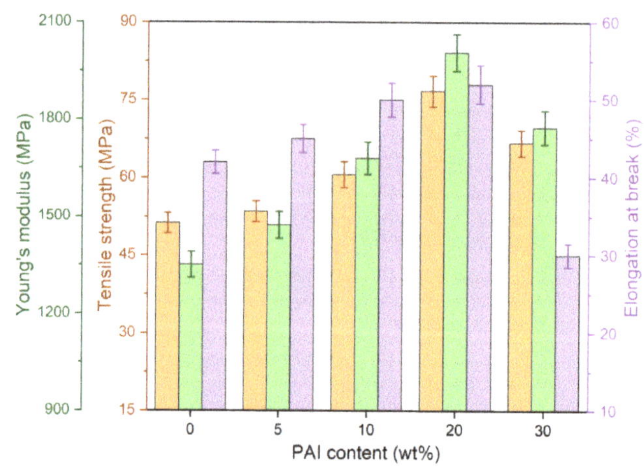

Figure 8. Tensile properties of PA6/PAI random copolymers.

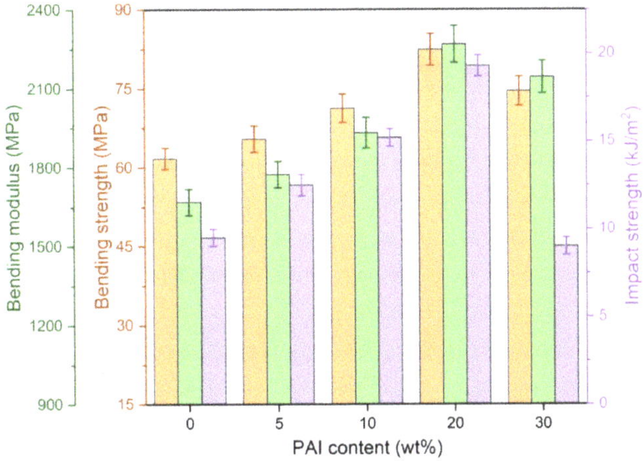

Figure 9. Bending and impact properties of PA6/PAI random copolymers.

3.6. Dielectric Properties

The dielectric properties of PA6/PAI random copolymers were evaluated by the dielectric constant and dielectric loss, as shown in Figure 10. The changes in the dielectric constant and dielectric of the PA6/PAI random copolymer could mainly be related to the number of orientable dipoles and the ability of dipole motion [32]. With the increase in the PAI content, the dielectric constant and dielectric loss of the PA6/PAI random copolymer decreased first and then increased. When the content of PAI was in the range of 0–20 wt%, the increasing aromatic imide groups improved the rigidity of the chain segment and impeded the motion of the dipole, which was not conducive to the polarization of the dipole but led to the reduction in the dielectric constant and dielectric loss. The dielectric constant and dielectric loss of PA6/PAI-20 markedly decreased to 3.85 and 0.01 from 5.12 and 0.10 for PA6, respectively. After that, the dielectric constant and dielectric loss of the PA6/PAI random copolymer increased with the further increase in the PAI content. This was because the introduction of excessive aromatic imide groups seriously destroyed the crystallization of the copolymer and increased the number of orientable dipoles, thus increasing the dielectric constant and dielectric loss.

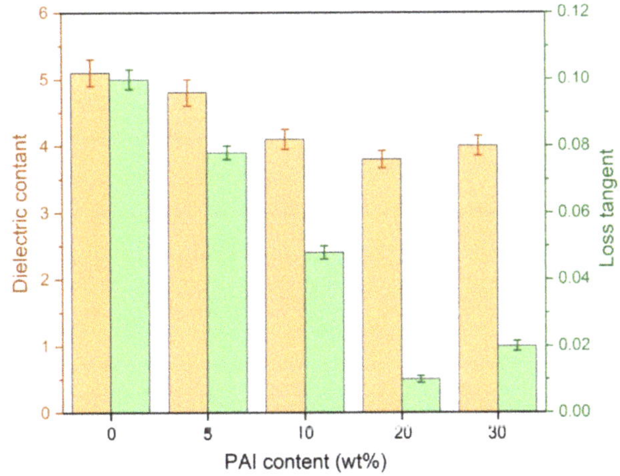

Figure 10. Dielectric constants and loss tangents of PA6/PAI random copolymers.

4. Conclusions

In this work, the DIDA monomer was obtained by the dehydration cyclization of PMDA and 6-aminocaproic acid, which was then reacted with HMDA to form PAI salt. PA6/PAI random copolymers with a different PAI content were successfully synthesized via the random copolymerization of CL and AI salt, and their thermal properties, crystallization, rheological behaviors, and mechanical and dielectric properties were systematically investigated. The results indicate that PA6/PAI random copolymers containing aromatic imide structure exhibit excellent thermal stabilities, and the T_d value and char yield increase obviously with the increase in the PAI content. The introduction of the aromatic imide structure did not change the crystal form but led to a decrease in crystallinity. As the proportion of PAI increases, the T_c and T_m values of the PA6/PAI random copolymer gradually decrease, but its T_g value increased obviously. Compared with pure PA6, the η^* values of PA6/PAI random copolymers were higher, but they still had good processabilities because of their lower T_m values. The introduction of an aromatic imide structural unit into the PA6 chain had a great influence on their mechanical and dielectric properties. When the PAI content was 20 wt%, the copolymer PA6/PAI-20 not only had high strength, modulus, and toughness but also demonstrated a low dielectric constant and dielectric loss. These results indicate that this method was feasible for designing and synthesizing new PA6 copolymers with high thermal stabilities and excellent mechanical and dielectric properties.

Author Contributions: Conceptualization, Y.Z., W.W. and C.W.; investigation, Y.Z. and Y.Y.; validation, Y.Z. and Y.Y.; writing—original draft preparation, Y.Z. and C.W.; writing—review and editing, C.W. and J.Y.; supervision, W.W. and J.Y. All authors have read and agreed to the published version of the manuscript.

Funding: This research was funded by the Science and Technology Innovation Program of Hunan Province, grant number 2021RC4072.

Institutional Review Board Statement: Not applicable.

Informed Consent Statement: Not applicable.

Data Availability Statement: Not applicable.

Conflicts of Interest: The authors declare no conflict of interest.

References

1. Goh, P.S.; Ismail, A.F. Chemically functionalized polyamide thin film composite membranes: The art of chemistry. *Desalination* **2020**, *495*, 114655. [CrossRef]
2. Pervaiz, M.; Faruq, M.; Jawaid, M.; Sain, M. Polyamides: Developments and applications towards next-generation engineered plastics. *Curr. Org. Synth.* **2017**, *14*, 146–155. [CrossRef]
3. Varghese, M.; Grinstaff, M.W. Beyond nylon 6: Polyamides via ring opening polymerization of designer lactam monomers for biomedical applications. *Chem. Soc. Rev.* **2022**, *51*, 8258–8275. [CrossRef] [PubMed]
4. Wang, C.H.; Zhang, Y.W.; Yi, Y.; Lai, D.W.; Yang, J.; Wang, W.Z. Thermal, morphological and mechanical properties of glass fiber reinforced star-branched polyamide 6. *Polym. Composite.* **2022**, *43*, 1617–1625. [CrossRef]
5. Fu, X.B.; Yao, C.G.; Yang, G.S. Recent advances in graphene/polyamide 6 composites: A review. *RSC Adv.* **2015**, *5*, 61688–61702. [CrossRef]
6. Zierdt, P.; Mitzner, E.; Gomoll, A.; Theumer, T.; Lieske, A. Synthesis of polyamide 6/11 copolymers and their use as matrix polymer in wood-plastic composites. *J. Appl. Polym. Sci.* **2016**, *133*, 44155. [CrossRef]
7. Bertolla, M.; Cecchetto, M.; Comotto, M.; Dal Moro, A. Comparison of the properties of a random copolymer and a molten blend PA6/PA6.9. *Polymers* **2022**, *14*, 4115. [CrossRef]
8. Cakir, S.; Eriksson, M.; Martinelle, M.; Koning, C.E. Multiblock copolymers of polyamide 6 and diepoxy propylene adipate obtained by solid state polymerization. *Eur. Polym. J.* **2016**, *79*, 13–22. [CrossRef]
9. Yang, Y.Y.; Kong, W.B.; Yuan, Y.; Zhou, C.L.; Cai, X.F. Structure-properties relationships of novel poly(carbonate-co-amide) segmented copolymers with polyamide-6 as hard segments and polycarbonate as soft segments. *J. Mol. Struct.* **2018**, *1157*, 52–60. [CrossRef]
10. Ai, T.H.; Zou, G.J.; Feng, W.T.; Ren, Z.L.; Li, F.; Wang, P.L.; Lu, B.; Ji, J.H. Synthesis and properties of biobased copolyamides based on polyamide 10T and polyamide 56 through one-pot polymerization. *New J. Chem.* **2021**, *45*, 14677–14686. [CrossRef]

11. Xu, L.; Li, J.J.; Wang, D.L.; Huang, Y.J.; Chen, M.M.; Li, L.B.; Pan, G.Q. Structure of polyamide 6 and poly (p-benzamide) in their rod-coil-rod triblock copolymers investigated with in situ wide angle X-ray diffraction. *Polymer* **2011**, *52*, 1197–1205. [CrossRef]
12. Wu, Z.Y.; He, J.J.; Yang, H.X.; Yang, S.Y. Progress in aromatic polyimide films for electronic applications: Preparation, structure and properties. *Polymers* **2022**, *14*, 1269. [CrossRef] [PubMed]
13. Ji, D.Y.; Li, T.; Hu, W.P.; Fuchs, H. Recent progress in aromatic polyimide dielectrics for organic electronic devices and circuits. *Adv. Mater.* **2019**, *31*, 1806070. [CrossRef] [PubMed]
14. Liaw, D.J.; Wang, K.L.; Huang, Y.C.; Lee, K.R.; Lai, J.Y.; Ha, C.S. Advanced polyimide materials: Syntheses, physical properties and applications. *Prog. Polym. Sci.* **2012**, *37*, 907–974. [CrossRef]
15. Yang, Z.H.; Ma, P.C.; Li, F.R.; Guo, H.Q.; Kang, C.Q.; Gao, L.X. Ultrahigh thermal-stability polyimides with low CTE and required flexibility by formation of hydrogen bonds between poly (amic acid)s. *Eur. Polym. J.* **2021**, *148*, 110369. [CrossRef]
16. He, J.J.; Yang, H.X.; Zheng, F.; Yang, S.Y. Dielectric properties of fluorinated aromatic polyimide films with rigid polymer backbones. *Polymers* **2022**, *14*, 649. [CrossRef] [PubMed]
17. Carriere, F.J.; Sekiguchi, H.; Surin, N.N.; Koteinikov, V.A.; Vygodskii, Y.S. Synthesis and thermal properties of polyamide 6 (A)-polyimide (B) block copolymers of ABA type. *Polym. Bull.* **1995**, *35*, 441–448. [CrossRef]
18. Ding, H.; Harris, F.W. Polyimide-nylon 6 copolymers: Single-component molecular composites. *Nanotechnology* **1996**, *622*, 273–291.
19. Pae, Y.; Harris, F.W. Synthesis and properties of novel polyimide/nylon-6 triblock copolymers. *J. Polym. Sci. Pol. Chem.* **2000**, *38*, 4247–4257. [CrossRef]
20. Pae, Y. Preparation of polyimide/nylon 6 graft copolymers from polyimides containing ester moieties: Synthesis and characterization. *J. Appl. Polym. Sci.* **2006**, *99*, 309–318. [CrossRef]
21. Pae, Y. Structure and properties of polyimide-g-nylon 6 and nylon 6-b-polyimide-b-nylon 6 copolymers. *J. Appl. Polym. Sci.* **2006**, *99*, 300–308. [CrossRef]
22. Zhang, Y.W.; Wang, C.H.; Yi, Y.; Wang, W.Z.; Yang, J.; Lou, Y.B.; Li, J.L. Synthesis and characterization of bio-based poly (amide imide)s derived from 11-aminoundecanoic acid and 1,10-diaminodecane. *J. Biobased Mater. Bio.* **2022**, *16*, 367–373. [CrossRef]
23. Wang, C.H.; Hu, F.; Yang, K.J.; Hu, T.H.; Wang, W.Z.; Deng, R.S.; Jiang, Q.B.; Zhang, H.L. Synthesis and properties of star-branched nylon 6 with hexafunctional cyclotriphosphazene core. *RSC Adv.* **2015**, *5*, 88382–88391. [CrossRef]
24. Wang, Y.B.; Yu, H.; Lu, G.M.; Luo, Z.L.; Shao, N.N.; Cao, M.; Wang, B.B. Influence of structural modification on the properties of poly (amide-imide)s. *Polym. Polym. Compos.* **2020**, *28*, 26–34. [CrossRef]
25. Wang, Y.B.; Lu, G.M.; Wang, W.J.; Cao, M.; Luo, Z.L.; Shao, N.N.; Wang, B.B. Molecular design and synthesis of thermotropic liquid crystalline poly (amide imide)s with high thermal stability and solubility. *E Polym.* **2017**, *17*, 199–207. [CrossRef]
26. Gechele, G.B.; Mattiussi, A. Intrinsic viscosity-molecular weight relations for hydrolytic polycaprolactam. *Eur. Polym. J.* **1965**, *1*, 47–61. [CrossRef]
27. Xu, Z.; Gao, C. In situ polymerization approach to graphene-reinforced nylon-6 composites. *Macromolecules* **2010**, *43*, 6716–6723. [CrossRef]
28. Shi, K.H.; Ye, L.; Li, G.X. Thermal oxidative aging behavior and stabilizing mechanism of highly oriented polyamide 6. *J. Therm. Anal. Calorim.* **2016**, *126*, 795–805. [CrossRef]
29. Kim, H.; Oh, K.; Seo, Y. Rheological and mechanical properties of a novel polyamide 6 synthesized by anionic polymerization of ε-caprolactam in a twin-screw extruder. *Polymer* **2019**, *177*, 196–201. [CrossRef]
30. Cao, L.S.; Jiang, T.H.; Liu, B.J.; Li, M.; Zhang, D.; Gong, W.; He, L. Effects of grafting and long-chain branching structures on rheological behavior, crystallization properties, foaming performance, and mechanical properties of polyamide 6. *e-Polymers* **2022**, *22*, 249–263. [CrossRef]
31. Qu, C.Y.; Tang, Y.; Wang, D.Z.; Fan, X.P.; Li, H.F.; Liu, C.W.; Su, K.; Zhao, D.X.; Jing, J.Q.; Zhang, X. Improved processability of PA66-polyimide copolymers with different polyimide contents. *J. Appl. Polym. Sci.* **2020**, *138*, 49640. [CrossRef]
32. Ning, J.Y.; Tian, C.; Yang, Y.Y.; Huang, L.; Lv, J.H.; Zeng, F.H.; Liu, Q.; Zhao, F.Q.; Kong, W.B.; Cai, X.F. A novel intrinsic semi-aromatic polyamide dielectric toward excellent thermal stability, mechanical robustness and dielectric performance. *Polymer* **2021**, *234*, 124233. [CrossRef]

polymers

Article

The Investigation of Copolymer Composition Sequence on Non-Isothermal Crystallization Kinetics of Bio-Based Polyamide 56/512

Diansong Gan [1,2], Yuejun Liu [1,*], Tianhui Hu [2], Shuhong Fan [1], Xiaochao Liu [1], Lingna Cui [1], Ling Yang [1], Yincai Wu [1], Lily Chen [1] and Zhixiang Mo [1]

1 Key Laboratory of Advanced Packaging Materials and Technology of Hunan Province, School of Packaging and Materials Engineering, Hunan University of Technology, Zhuzhou 412007, China; gands@zztep.cn (D.G.); f1s1h1@163.com (S.F.); xcliu_2014@163.com (X.L.); lncui1102@126.com (L.C.); yangling00002@163.com (L.Y.); xmwuyincai@fjirsm.ac.cn (Y.W.); lily_jasmine_001@163.com (L.C.); mzx13873355623@163.com (Z.M.)

2 Zhuzhou Times Engineering Plastics Industrial Co., Ltd., Zhuzhou 412008, China; huth@zztep.cn

* Correspondence: yjliu_2005@126.com

Abstract: A new bio-based polyamide 56/512 (PA56/512) has been synthesized with a higher bio-based composition compared to industrialized bio-based PA56, which is considered a lower carbon emission bio-based nylon. In this paper, the one-step approach of copolymerizing PA56 units with PA512 units using melt polymerization has been investigated. The structure of the copolymer PA56/512 was characterized using Fourier-transform infrared spectroscopy (FTIR) and Proton nuclear magnetic resonance (^{1}H NMR). Other measurement methods, including relative viscosity tests, amine end group measurement, thermogravimetric analysis (TGA) and differential scanning calorimetry (DSC), were used to analyze the physical and thermal properties of the PA56/512. Furthermore, the non-isothermal crystallization behaviors of PA56/512 have been investigated with the analytical model of Mo's method and the Kissinger method. The melting point of copolymer PA56/512 exhibited a eutectic point at 60 mol% of 512 corresponding to the typical isodimorphism behavior, and the crystallization ability of PA56/512 also displayed a similar tendency.

Keywords: copolymer PA56/512; bio-based polymer; thermal properties; non-isothermal crystallization kinetics

Citation: Gan, D.; Liu, Y.; Hu, T.; Fan, S.; Liu, X.; Cui, L.; Yang, L.; Wu, Y.; Chen, L.; Mo, Z. The Investigation of Copolymer Composition Sequence on Non-Isothermal Crystallization Kinetics of Bio-Based Polyamide 56/512. *Polymers* **2023**, *15*, 2345. https://doi.org/10.3390/polym15102345

Academic Editors: Hui Zhao, Wei Wu and Bin Wang

Received: 16 April 2023
Revised: 13 May 2023
Accepted: 15 May 2023
Published: 17 May 2023

1. Introduction

Polyamides (PAs) have displayed versatile potential in the fabrication of aluminum soft packaging film, especially for lithium-polymer batteries. PA also offers advantages such as high thermal stability, excellent chemical resistance and superior mechanical performance due to its unique amide group structures and arrangement of hydrogen bonds [1–3]. The aluminum–plastic film is a key part of lithium-polymer batteries, contributing to the safety, long-term stability and whole service life time of the battery itself [4]. The biaxially oriented polyamide (BOPA) film made of PA6 is commonly used as the outer layer of aluminum–plastic film. It requires good impact resistance, puncture resistance, insulation performance and friction resistance to protect the inner layer structure of the battery during normal usage [5]. With the growth of the NEV market, the use of eco-friendly polyamide films has been a hot topic, in line with the new global low-carbon emission strategy and trend.

The cost-effective industrial production of bio-based 1,5-pentanediamine from Cathay Biotech Inc. has attracted increased research for the development of bio-based polyamides. The bio-based 1,5-pentanediamine is obtained by the decarboxylation of lysine produced by the biological fermentation of natural raw materials such as starch and straw [6,7]. Bio-based polyamides with the required characteristics and structure, such as long-chain polyamides, high-temperature polyamides and transparent polyamides, can be produced using polycondensation processes with dioic acids of different chain lengths [8–13]. The commercialized

PA56, polymerized from bio-based 1,5-pentanediamine and petroleum-based adipic acid, is a promising matrix film resin for use in the next generation of aluminum soft packaging film with green chemical energy applications in lithium–polymer batteries.

The low toughness and high water absorption capacity of pure PA56 would negatively affect the mechanical properties, electrical properties and dimensional stability of the products. To address this issue, a compounding modification and copolymerization process has often been used for the improvement of the performance of PA56 [14,15]. Recent work by Xu et al. [16] has demonstrated that the blend of PA56 and EPDM-g-MAH can significantly improve the toughness and heat resistance of the system. Yan et al. [17] proposed that PA56 could be used as the internal lubricating and nucleating agent in a blend system of PA56 and polyethylene terephthalate. Yang et al. [18] reported that the inherently flame-retardant PA56 copolymers (FRPA56s) synthesized by melt polycondensation with monomer 1,5-pentanediamine (PDA), adipic acid (AA) and 9,10-dihydro-10-[2,3-di(hydroxy carbonyl)propyl]-10-phosphaphenanthrene-10-oxide (DDP) presented an improvement in the flame retarding performance. Studies on the synthesis and performance of copolymer PA5T/56 [19], PA10T/56 [20] and PA56/66 [21] have also been reported before, although there are still few studies on the modification of PA56 through the copolymerization of long carbon chain aliphatic polyamide monomers. The increasing length of the methylene segment of the copolymer polyamide would increase the possibilities for hydrogen bonding arrangements, conformational transitions and structure changes in the system [22].

This paper demonstrates a high-efficiency approach to copolymerizing PA56 units with PA512 units through melt polymerization. The relative viscosity and the amine end group values of the samples were measured using an Ubbelohde viscometer and Metrohm 888 Titrando, respectively. The structure of the synthesized PA56/512 was characterized by FTIR and ^{1}H NMR, and the thermal properties of PA56/512 were measured by thermogravimetric analysis (TGA) and differential scanning calorimetry (DSC). The long carbon chain of PA512 allows its methylene group to undergo more flexible conformational transition in the co-polymer component, resulting in different non-isothermal crystallization behaviors. Furthermore, the non-isothermal crystallization behaviors of PA56/512 with different composition ratios at different cooling rates were studied based on the DSC data. Mo's method and the Kissinger method were used to determine the crystallization behavior of PA56/512. In summary, this study presents a comprehensive analysis of the copolymerization of PA56 units with PA512 units through melt polymerization. The results provide valuable insights into the thermal and crystallization properties of the synthesized PA56/512 copolymer, which can be useful in various industrial applications.

2. Experimental Section

2.1. Materials

1,5-pentanediamine was purchased from Cathay Biotech Inc. (Shanghai, China). Adipic acid was purchased from Sinopec Yangzi Petrochemical Co. Ltd. (Jiangsu, China). Dodecanedioic acid, concentrated sulfuric acid (96%), trifluoroethanol and trifluoroacetic acid-d (isotopic) were purchased from Aladdin Industrial Corporation (Shanghai, China). All reagents used were of analytically pure quality. All solutions were prepared using deionized water.

2.2. Preparation of Copolymer PA56/512

2.2.1. Neutralization Reaction

The adipic acid and dodecanedioic acid, with a total weight of 2.5 kg, were prepared in a 50% mixed aqueous solution according to the set molar ratio (8:2, 7:3, 6:4, 5:5, 4:6, 3:7, 2:8). Then, 1,5-pentanediamine was added dropwise to the above mixed aqueous solution under vigorous stirring at 80 °C. A transparent aqueous solution was obtained when the pH of the whole solution was regulated at 7.8.

2.2.2. Polymerization Reaction

The above solution was transferred to a 10 L autoclave, and the air in the autoclave was replaced with nitrogen 3–5 times. The reactor was heated to 155 °C to remove 1.1 kg of water; the temperature was increased to 200 °C and the pressure was maintained at 1.5 MPa for an hour. After the autoclave was heated to 260 °C with 1.5 MPa pressure, the vapor in the autoclave was gradually released to reduce the pressure to normal. The system was vacuumed and reacted at −0.08 MPa for 1 h. Finally, the PA56/512 resin was obtained after the discharging, cooling and granulating process.

Besides this, pure PA56 and pure PA512 resins were prepared in the same way, named sample #1 and #9, respectively. The copolymers PA56/512 were prepared from monomeric adipic acid and dodecanedioic acid with ratios of 8:2, 7:3, 6:4, 5:5, 4:6, 3:7 and 2:8, named samples #2, #3, #4, #5, #6, #7 and #8, respectively. Figure 1 presents a schematic diagram of the polymerization process of PA56/512, as well as the structure of pure PA56 and pure PA512.

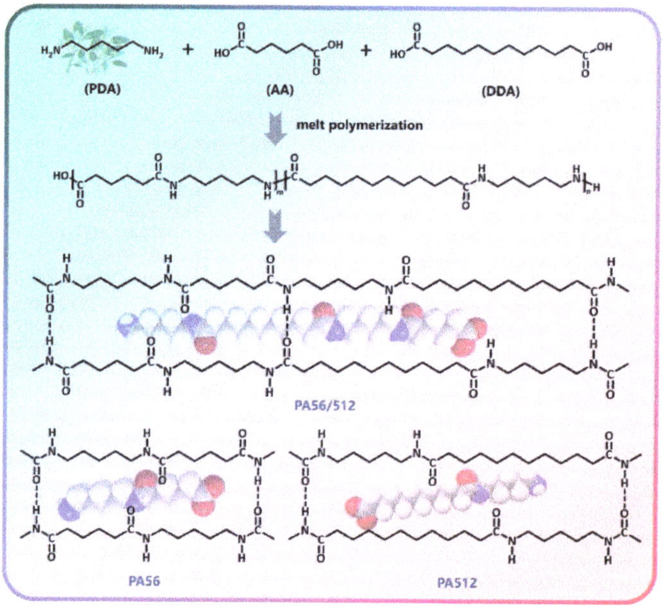

Figure 1. Schematic diagram of the polymerization process of PA56/512.

2.3. Characterization of Copolymer PA56/512

Fourier transform infrared spectrometry (FTIR, PerkinElmer Frontier, Waltham, MA, USA) was carried out to obtain the functional group information of copolymer PA56/512. ^{1}H NMR was measured using a Bruker (Billerica, MA, USA) AVIII 500HD (500 MHz) instrument with a solvent of trifluoroacetic acid to analyze the chemical structures of samples. The relative viscosities of the samples with a concentration of 0.5 g/mL were measured using an Ubbelohde viscometer in sulfuric acid solution (96%) at 25 °C. The concentrations of the amine end group were determined using Metrohm 888 Titrando (Herisau, Switzerland). In total, 0.1 g PA56/512 was fully dissolved in 50 mL trifluoroethanol solution, which was titrated using the hydrochloric acid–ethanol solution to measure the concentrations of amine end groups. Thermogravimetric analyses (TGA/DSC1/1100SF, METTLER TOLEDO, Columbus, OH, USA) of the samples were performed in N_2 flow with a heating increase of 10 °C/min from 25 °C to 700 °C. The thermal performance and the non-isothermal crystallization kinetics of the prepared samples were measured by differential scanning calorimetry (DSC 214, NETZSCH, Selb, Germany). All samples were heated from 25 °C to 280 °C, with a heating rate of 10 °C/min. After holding at 280 °C for 3 min, the program was set to cool down to 25 °C at the set cooling rates of 2 °C/min, 5 °C/min, 10 °C/min and 20 °C/min respectively. After holding at 25 °C for 3 min, the second heating program was carried out in the same way as the first. All the DSC tests above were conducted in a nitrogen atmosphere.

3. Results and Discussion

3.1. Structural Characterization of PA56/512

Figure 2 shows the FTIR spectra of PA56/512 with different composition ratios. The similar peak areas obeserved in samples #1–9 demonstrate that all the samples contained the same functional group. The stretching vibrations of N–H bonds can be seen at the wavenumber of 3300 cm^{-1}. The characteristic peaks observed at 2927 cm^{-1} and 2857 cm^{-1} correspond to the asymmetric stretching vibration and symmetric stretching vibration of $-CH_2$, respectively. The sharp peak overserved at 1636 cm^{-1} belonging to Amide I resulted from the C=O stretching vibration. The two peaks in Amide II and Amide III, corresponding to the wavenumbers of 1540 cm^{-1} and 1270 cm^{-1} respectively, are attributed to the mixing interaction of N–H bending vibrations and C–N stretching vibrations. Additionally, the 930 cm^{-1} (Amide IV) and 690 cm^{-1} (Amide V) peaks represent the C–C=O stretching vibration and N–H out-of-plane wagging vibration, respectively [23]. The presence of Amide I–V peaks indicates that all the PA56/512 samples (#1–9) prepared have a typical polyamide structure.

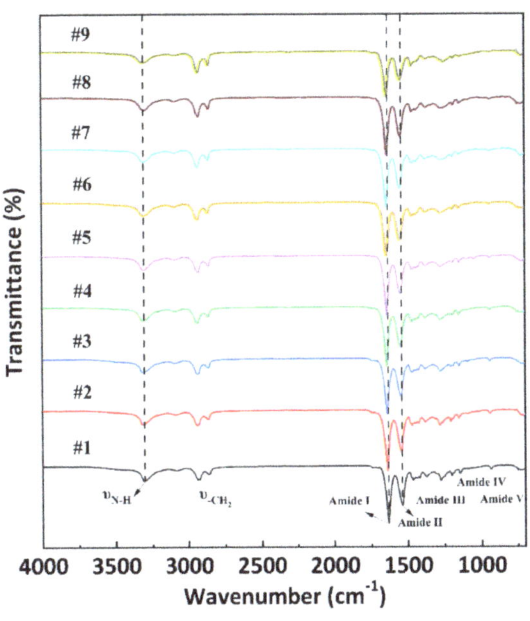

Figure 2. FTIR spectra of PA56/512 with different composition ratios.

The ^{1}H NMR spectra of PA56/512 with different composition ratios are shown in Figure 3. Regions #a (3.53–3.35 ppm) and #b (2.68–2.56 ppm) represent the characteristic peak positions of α–H (α_2, α_4) of diamines and α–H (α_1, α_3) of diacids in PA56/512, respectively. Likewise, all the β–H (β_1, β_2, β_3, β_4) of diamines and diacids in PA56/512 are displayed in region #c (1.85–1.56 ppm), while the remaining γ/δ/ε–H (γ_2, γ_3, γ_4, δ_3, ε_3) appear in region #d (1.45–1.21 ppm). Furthermore, the ratios of integral data in each region of the ^{1}H NMR spectra are listed in Table 1, including the theoretical and experimental data. It is evident that the experimental values are consistent with the theoretical value, indicating the successful preparation of PA56/512 with different composition ratios. Furthermore, the relative viscosities and amine end group values of PA56/512 are also listed in Table 1. It can be clearly seen that the relative viscosity and M_n values of PA56/512 for all ratios are above 2.4 and 19,000 respectively, revealing effective one-step polymerization in the copolymerization of PA56 and PA512.

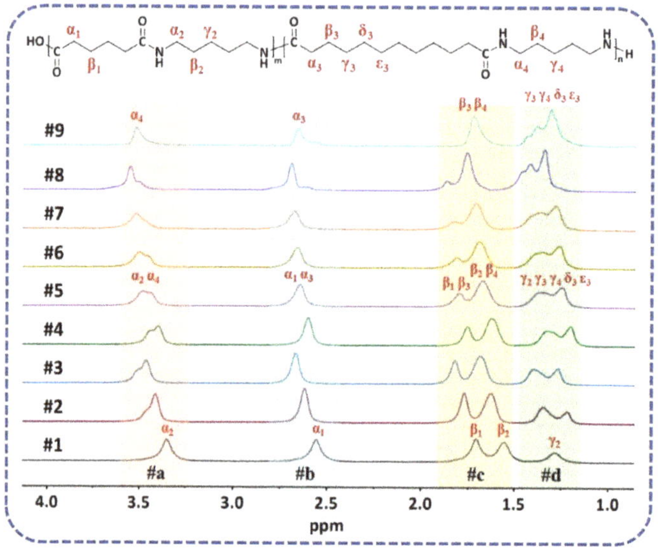

Figure 3. ^{1}H NMR spectra of PA56/512 with different composition ratios.

Table 1. The ratios of integral data in each region of the ^{1}H NMR spectra and other relevant values for PA56/512.

Sample	$S_{\#a}$:$S_{\#b}$:$S_{\#c}$:$S_{\#d}$ Theoretical Value	$S_{\#a}$:$S_{\#b}$:$S_{\#c}$:$S_{\#d}$ Experimental Value	Relative Viscosity	$[NH_2]$ (mol/t)	$M_n-[NH_2]$ (g/mol)
#1	1:1:2:0.5	1:1:2:0.53	2.89	38.4	26,041.7
#2	1:1:2:1.1	1:1:2:1.04	2.82	41.3	24,213.1
#3	1:1:2:1.4	1:1:2:1.33	2.71	45.7	21,881.8
#4	1:1:2:1.7	1:1:2:1.63	2.69	46.4	21,551.7
#5	1:1:2:2.0	1:1:2:1.94	2.60	48.2	20,746.9
#6	1:1:2:2.3	1:1:2:2.25	2.53	51.2	19,531.3
#7	1:1:2:2.6	1:1:2:2.59	2.59	48.5	20,618.6
#8	1:1:2:2.9	1:1:2:2.82	2.55	49.1	20,366.6
#9	1:1:2:3.5	1:1:2:3.54	2.46	50.4	19,841.3

3.2. Thermal Properties of PA56/512

Figure 4a,b illustrate the TGA and DTG variation curves of PA56/512 with different composition ratios used to measure the thermal stability of all the samples. The relevant data on the process of thermal decomposition are listed in Table 2. It is clear that pure PA56 and PA512, and their copolymers, exhibit the same one-step thermal decomposition process, which is consistent with the typical thermal degradation mechanism of polyamides [24,25]. In general, C–N bond breakage occurs first, accompanied by the cyclization of pentanediamine, the hydrolysis of chain-segments and other reactions, and then C–C bond breakage occurs at high temperature [26]. According to the thermal decomposition data shown in Table 2, the initial decomposition temperature (T_D), 20% weight decomposition temperature ($T_{20\%}$) and maximum decomposition rate temperature (T_M) mostly increased with the increase in 512 content in PA56/512, revealing the better thermal stability of pure PA512 compared to pure PA56, resulting from the lower hydrogen bond density of PA512 than that of PA56 [27].

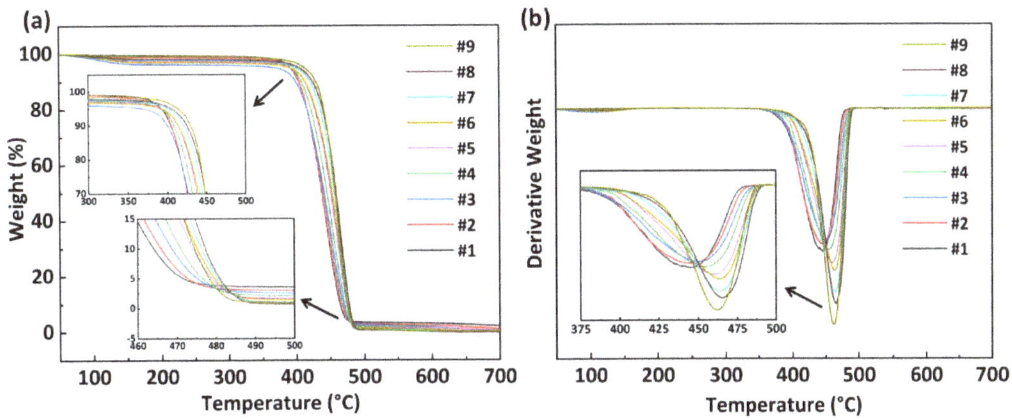

Figure 4. (**a**) TGA and (**b**) DTG variation curves of PA56/512 with different composition ratios.

Table 2. Thermal decomposition data of PA56/512 with different composition ratios from TGA.

Sample	T_D/°C	$T_{20\%}$/°C	T_M/°C
#1	277.1	416.5	444.6
#2	281.8	416.7	445.6
#3	284.8	416.9	450.4
#4	297.6	422.6	454.9
#5	299.2	428.2	460.4
#6	309.4	429.7	463.6
#7	311.4	436.4	464.7
#8	318.8	440.1	466.2
#9	316.4	441.6	462.6

The non-isothermal crystallization curves of PA56/512 with different composition ratios at different cooling rates are shown in Figure 5. It can be observed that the crystallization temperature of PA56/512 display a trend of first decreasing and then increasing with the increase in 512 content, accompanied by a peak shape of first widening and then narrowing, due to the crystallization competition between PA56 and PA512. Furthermore, the pattern of the above trend is consistent at different cooling rates, and the second heating cycle DSC curves of PA56/512 with different composition ratios at different cooling rates are shown in Figure 6. It can be clearly seen that the moving trend of the melting curve is analogous to that of the crystallization curve. Figures 6 and 7 depict the melting point curves of PA56/512 with different contents of 512. The copolymerization system of PA56/512 has the lowest melting point with 40% 56 and 60% 512, corresponding to sample #6, which exhibits typical isodimorphism behavior [28]. The lowest melting point is regarded as a "eutectic" point [29,30], which is considered as a demarcation point for crystallization behavior dominated by homopolymer crystal types of 56 or 512. Therefore, the preparation of random copolymer PA56/512 with different composition ratios is a simple and efficient means to obtain a rearranged sequence structure with tunable thermal properties. Moreover, the investigated system exhibits a wide range of tunable Tm, enabling a broad range of processing temperatures to be achieved, which could meet different industrial application requirements.

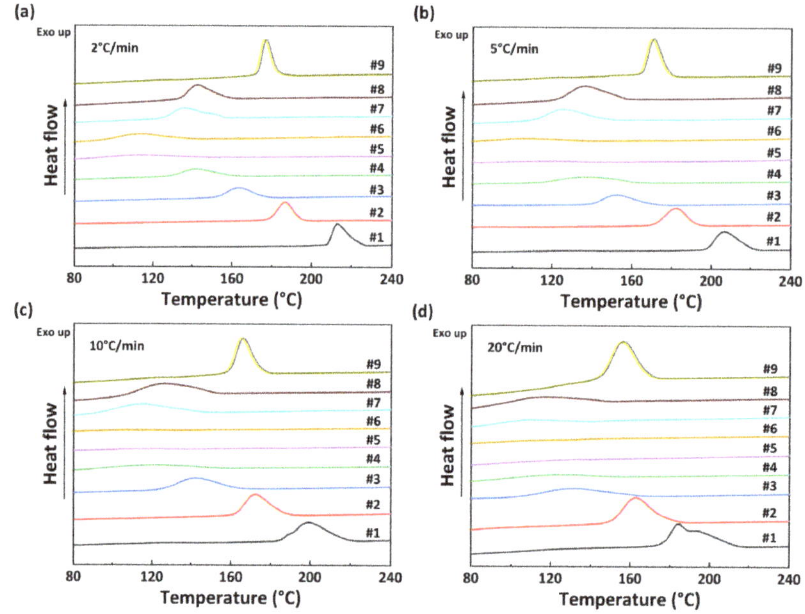

Figure 5. Non-isothermal crystallization curves of PA56/512 with different composition ratios at different cooling rates: (**a**) 2 °C/min, (**b**) 5 °C/min, (**c**) 10 °C/min, (**d**) 20 °C/min.

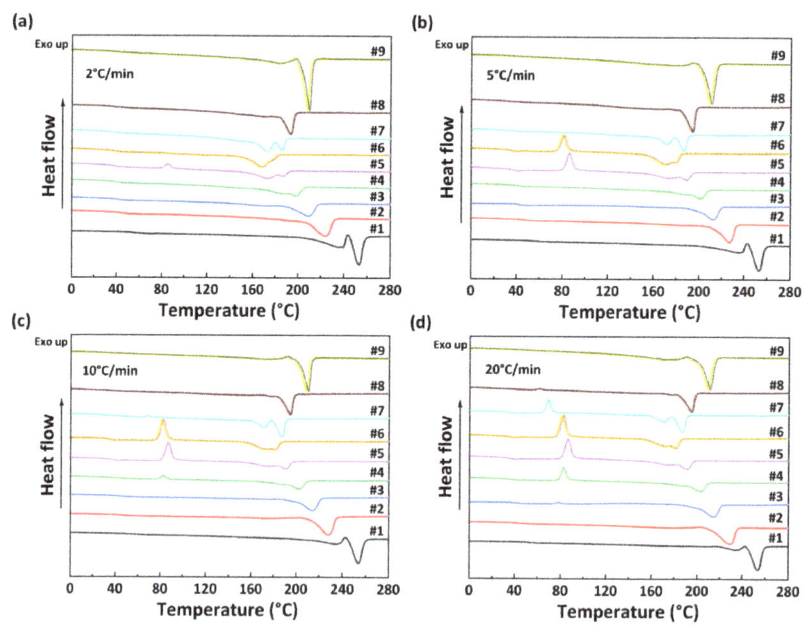

Figure 6. The second heating DSC curves of PA56/512 with different composition ratios at different cooling rates: (**a**) 2 °C/min, (**b**) 5 °C/min, (**c**) 10 °C/min, (**d**) 20 °C/min.

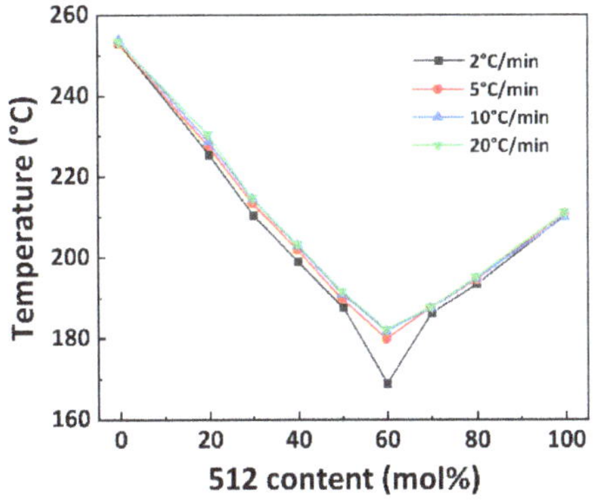

Figure 7. The melting temperature curve of PA56/512 with different contents of 512.

3.3. Non-Isothermal Crystallization Kinetics of PA56/512

Based on the crystallization curves of PA56/512 with different composition ratios, five sets of samples, #1, #2, #3, #8, and #9, are selected for further non-isothermal crystallization kinetics analysis. Among them, the crystallization behaviors of #1–3 and #8–9 are dominated by 56 and 512, respectively. Figure 8 illustrates the non-isothermal crystallization curves of PA56/512 at different cooling rates. It can be seen that as the cooling rate increases, the crystallization exothermic peaks of each sample shift to the lower temperatures, along with a gradually widening peak. This result is mainly attributed to the time difference between the rearrangement and stacking of polymer chain segments during the crystallization process [31,32]. Consequently, the greater the cooling rate, the more obvious the crystallization hysteresis. Otherwise, with a high cooling rate, the temperature of the samples decreases more significantly within the same cooling time, while the activity of the polymer molecular chains is poor at low temperatures, resulting in decreased crystallization perfection and a wider crystallization peak shape.

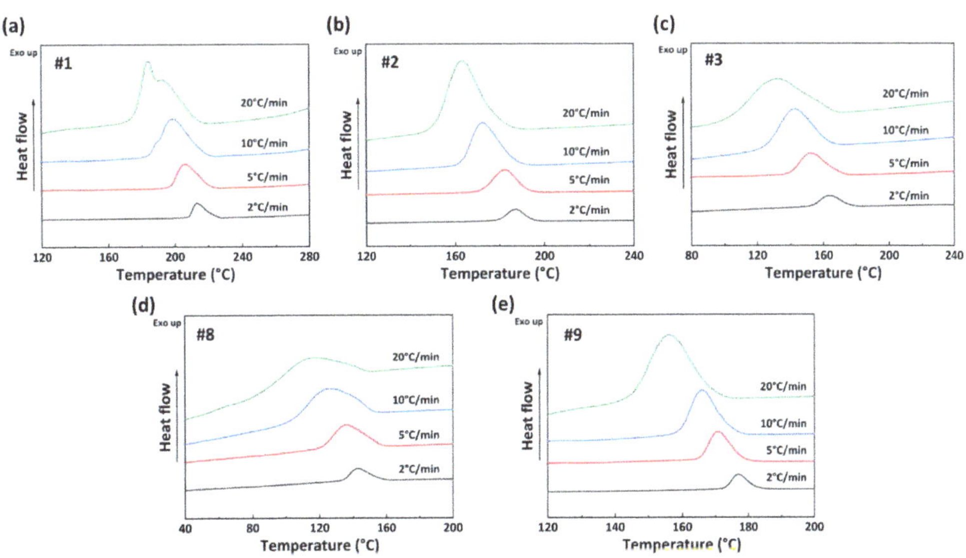

Figure 8. Non-isothermal crystallization curves of PA56/512 at different cooling rates: (**a**) #1, (**b**) #2, (**c**) #3, (**d**) #8, (**e**) #9.

The relative crystallinity *X(T)* of PA56/512 can be calculated according to the following formula [33]:

$$X(T) = \frac{\int_{T_0}^{T}\left(\frac{dH_c}{dT}\right)dT}{\int_{T_0}^{T_e}\left(\frac{dH_c}{dT}\right)dT} \tag{1}$$

where T_0 and T_e are the crystallization initiation temperature and crystallization termination temperature, respectively. dH_c/dT refers to the heat flow at a certain temperature. Figure 9 shows the *X(T)*−*T* curve of the PA56/512 obtained from Formula (1). The relevant non-isothermal crystallization kinetic parameters of PA56/512 with different composition ratios are listed in Table 3. The parameter Φ refers to cooling rates; T_p represents crystallization peak temperature; ΔT_c refers to the temperature interval between T_0 and T_e; ΔH_c is the enthalpy of crystallization; $t_{1/2}$ refers to the semi-crystalline period. According to Figure 9 and Table 3, at the same cooling rate, the crystallization temperature ranges of samples #1 to #3 increase gradually, while those of samples #8 to #9 decreases gradually, which corresponds to the crystallization times of samples #1 to #3 (#9 to #8) gradually becoming longer, and the crystallization rate gradually decreasing.

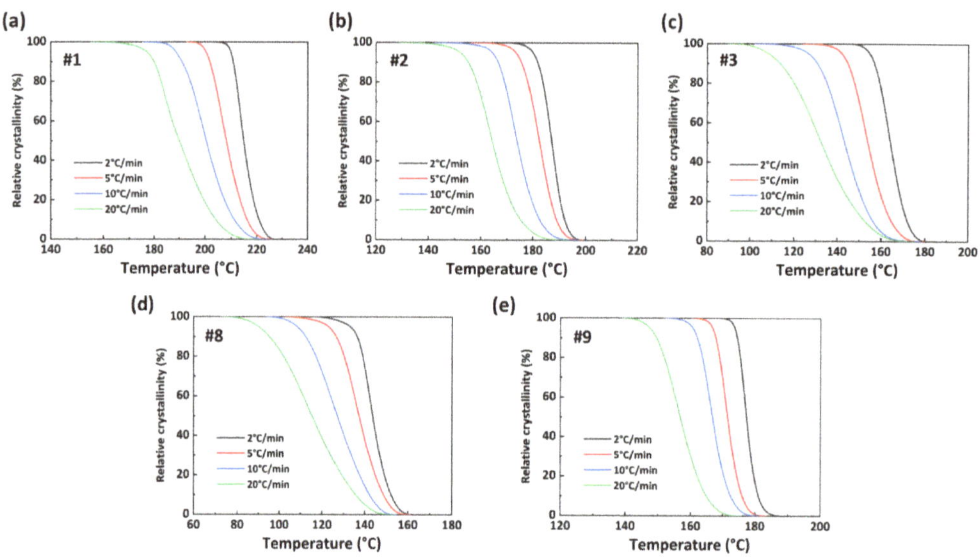

Figure 9. *X(T)*−*T* curves of PA56/512: (**a**) #1, (**b**) #2, (**c**) #3, (**d**) #8, (**e**) #9.

Table 3. Non-isothermal crystallization kinetic parameters of PA56/512 with different composition ratios.

Sample	Φ (°C/min)	T_0 (°C)	T_p (°C)	T_e (°C)	ΔT_c (°C)	ΔH_c (J/g)	$t_{1/2}$ (min)
#1	2	227.4	213.2	203.0	24.4	50.0	6.30
	5	225.9	206.6	193.0	32.9	50.0	3.56
	10	224.1	199.4	175.5	48.6	54.3	2.36
	20	219.8	185.0	155.3	64.5	49.4	1.50
#2	2	200.1	187.3	169.3	30.8	44.6	6.52
	5	198.7	182.6	156.2	42.5	43.5	3.25
	10	197.1	172.4	142.7	54.4	49.5	2.32
	20	193.1	163.5	128.9	64.2	49.9	1.44

Table 3. *Cont.*

Sample	Φ (°C/min)	T_0 (°C)	T_p (°C)	T_e (°C)	ΔT_c (°C)	ΔH_c (J/g)	$t_{1/2}$ (min)
#3	2	181.3	163.7	145.4	35.9	36.3	8.58
	5	179.7	152.5	124.1	55.6	37.1	5.17
	10	178.2	142.8	102.4	75.8	44.0	3.48
	20	175.1	130.7	89.8	85.3	33.8	2.13
#8	2	162.5	142.7	116.6	45.9	55.6	9.37
	5	161.2	135.7	102.7	58.5	60.5	4.81
	10	155.0	125.4	94.1	60.9	55.1	2.81
	20	153.3	114.9	72.8	80.5	33.6	1.90
#9	2	188.4	176.9	169.0	19.4	53.9	5.51
	5	183.4	171.1	160.4	23.0	54.9	2.39
	10	181.1	166.3	152.3	28.8	53.5	1.44
	20	175.6	156.1	139.2	36.4	57.9	0.94

The conversion relationship between crystallization time (t) and crystallization temperature (T) is shown in Equation (2) [32]:

$$t = \frac{|T_0 - T|}{\Phi} \tag{2}$$

Figure 10 presents the $X(t)-t$ curve of the PA56/512 obtained from Formulas (1) and (2). Combined with the results for the same sample in Figures 9 and 10, it can be noted that the crystallization rate of copolymer PA56/512 initially increases and then decreases at a specific cooling rate. The curve is relatively flat in the early and late crystallization stages, with a lower crystallization rate. This is mainly attributed to the fact that the nucleation process dominated by the nucleation rate requires a large degree of undercooling in the initial stage of crystallization, resulting in a lower nucleation rate [34]. In the middle of crystallization, both the nucleation rate and crystal growth rate need to be considered, resulting in the maximum crystallization rate when they match each other. Although the nucleation rate remains high in the late crystallization stage, the crystallization rate is dominated by the crystal growth rate [35]. The low temperature in this stage causes the movement of polymer chain segments to be blocked, thereby reducing the crystal growth rate. According to the data in Figure 10 and Table 3, the semi-crystallization time $t_{1/2}$ is ranked as #1 < #2 < #3 < #9 < #8, corroborating the above conclusion that the addition of 56 or 512 copolymerization components to the system will result in a lower crystallization rate than the original system (pure PA56 or PA512), similarly to the trend seen for melting point.

Mo et al. [36] combined the Avrami equation and the Ozawa equation to obtain a corresponding relationship with the same time and the same crystallinity. The equation is as follows [37]:

$$\lg\Phi = \lg F(T) - a\lg t \tag{3}$$

where the parameter a refers to the ratio of the Avrami index and the Ozawa index. The physical meaning of $F(T)$ can be interpreted as the cooling rate value that must be selected for a certain polymer system to reach a certain relative crystallinity per unit time [38]. Thus, the $F(T)$ value is inversely proportional to the crystallization rate.

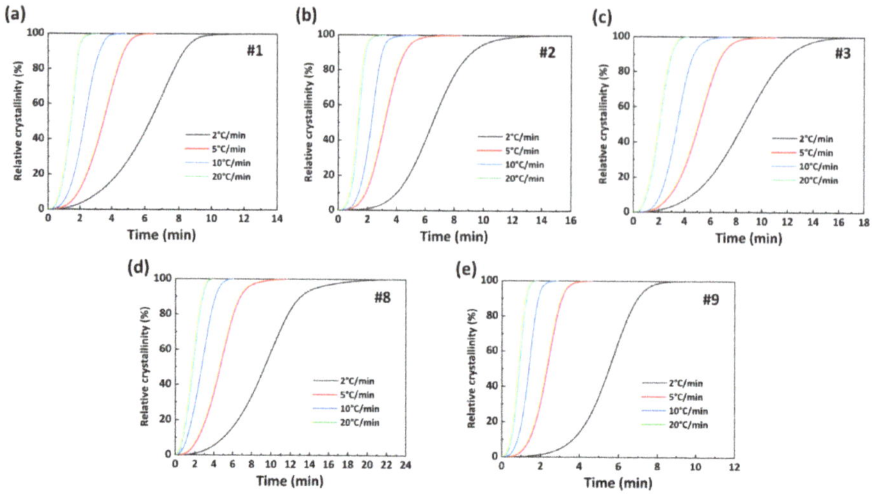

Figure 10. *X(t)−t* curves of PA56/512: (**a**) #1, (**b**) #2, (**c**) #3, (**d**) #8, (**e**) #9.

Based on Formula (3), the lgΦ−lg*t* curves of the PA56/512 with different composition ratios are shown in Figure 11. The high fitting degree of the equation to the curve reveals its applicability to the analysis of non-isothermal processes related to the PA56/512 system. The relevant calculation parameters are listed in Table 4. The value *a* of the same sample is approximately constant, and undergoes no change with differences in its own *X(t)*, confirming a certain proportional relationship between the Avrami index and Ozawa index for a specific system. The *F(T)* of the same sample is proportional to the relative crystallinity, indicating that the larger the relative crystallinity, the smaller the average crystallization rate in this range. Comparing the values of *F(T)* with different copolymer ratio systems with the same crystallinity, such as *X(t)* = 20%, the values of *F(T)* rank as follows: 21.18 (#1) < 22.32 (#2) < 41.45 (#3) < 11.47 (#9) < 23.56 (#8). The average crystallization rate exhibits an inverse relationship, which is consistent with the previous analysis. In addition, it can be seen that the average crystallization rate of pure PA512 is better than that of pure PA56.

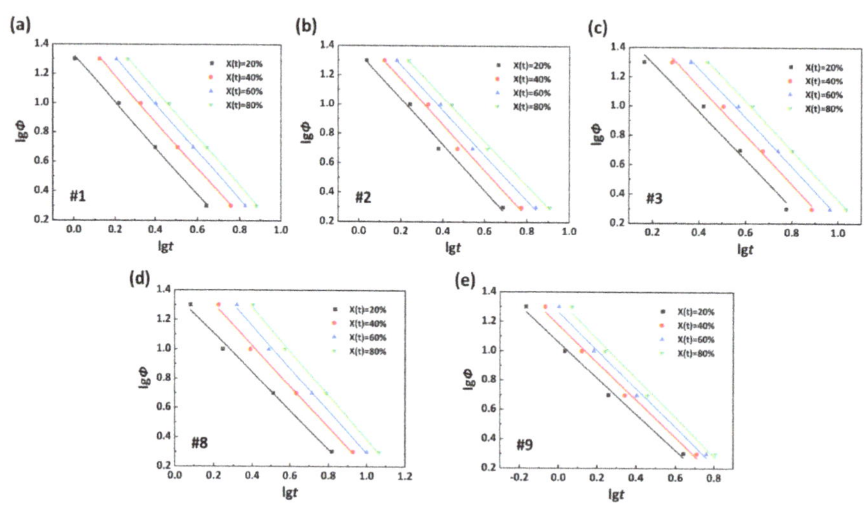

Figure 11. lgΦ−lg*t* curves of PA56/512: (**a**) #1, (**b**) #2, (**c**) #3, (**d**) #8, (**e**) #9.

Table 4. Non-isothermal crystallization kinetic parameters of PA56/512 based on the Mo method.

Sample	Parameter	X(t) = 20%	X(t) = 40%	X(t) = 60%	X(t) = 80%
#1	$F(T)$	21.18	32.64	44.82	56.52
	a	1.58	1.59	1.62	1.63
#2	$F(T)$	22.32	30.98	38.00	45.45
	a	1.56	1.55	1.53	1.50
#3	$F(T)$	41.45	64.06	85.15	111.98
	a	1.64	1.66	1.67	1.68
#8	$F(T)$	23.56	38.77	55.34	77.90
	a	1.32	1.39	1.45	1.50
#9	$F(T)$	11.47	15.26	18.64	22.76
	a	1.23	1.27	1.30	1.34

The activation energy of the non-isothermal crystallization of a polymer is often calculated using the equation of Kissinger, as follows [39]:

$$\frac{\mathrm{d}\left[\ln\frac{\Phi}{T_P{}^2}\right]}{\mathrm{d}\left(\frac{1}{T_p}\right)} = -\frac{\Delta E}{\mathrm{R}} \quad (4)$$

where T_p represents the crystallization peak temperature, ΔE refers to the activation energy, and R is the gas constant. The slope $-\Delta E/\mathrm{R}$ can be obtained from the linear relationship between $\ln(\Phi/T_p{}^2)$ and $T_p{}^{-1}$, and then the ΔE value can be calculated. The $\ln(\Phi/T_p{}^2)-T_p{}^{-1}$ curves of PA56/512 for samples #1, #2, #3, #8 and #9 are shown in Figure 12. The calculated crystallization activation energy data are listed in Table 5, along with the following rankings: #1 > #2 > #3 > #9 > #8 > #9 > #1. A high crystallization activation energy means high crystallization exotherm, accompanied by a strong molecular chain movement ability, a high crystallization rate and a strong crystallization ability [40]. Accordingly, pure PA512 has the best crystallization ability, followed by pure PA56. It can be inferred that the crystallization ability of copolymer PA56/512 is similar to the trend of its melting temperature, with the lowest level at the eutectic point, which conforms to all previous analyses of non-isothermal crystallization kinetics.

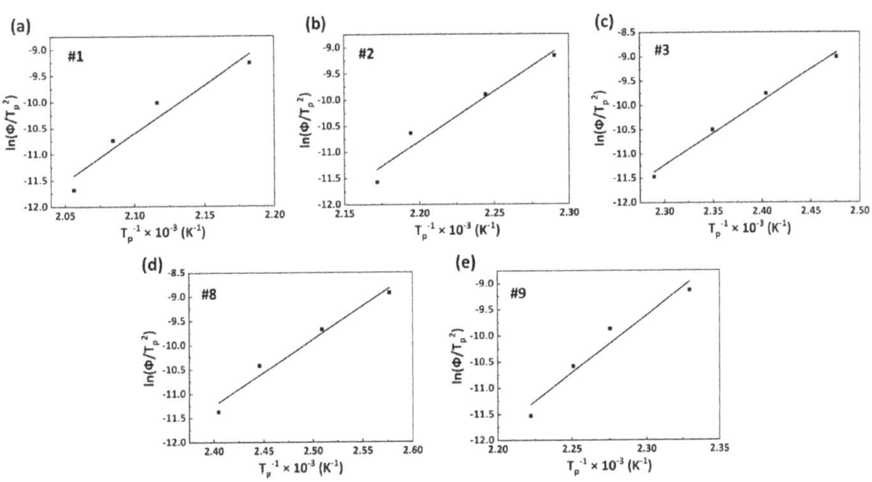

Figure 12. $\ln(\Phi/T_p{}^2)-T_p{}^{-1}$ curves of PA56/512: (**a**) #1, (**b**) #2, (**c**) #3, (**d**) #8, (**e**) #9.

Table 5. The crystallization activation energy of PA56/512.

Sample	#1	#2	#3	#8	#9
ΔE (kJ/mol)	156.6	154.2	109.1	113.6	181.5

In general, the reconfigurable chain structure and tunable thermal properties of random copolymer PA56/512 can be realized by regulating the ratio of copolymerized components, resulting in a difference in the arrangement of hydrogen bonds, which affects the crystallization process of the system, accompanied by a change in crystallization kinetics. Furthermore, all studies demonstrate that the average crystallization rate of pure PA512 is higher than that of PA56, owing to the longer saturated carbon chain of PA512, the greater mobility of the long polymethylene segments, and the more flexible orientation of hydrogen bonds [41,42].

4. Conclusions

In this paper, a bio-based copolymer PA56/512 with different composition ratios was successfully synthesized by a one-step melt polymerization process. All the samples exhibited high relative viscosity and molecular weight, which makes them suitable for use in industrial application processes. The FTIR and ^{1}H NMR spectra show that the actual structure of the prepared PA56/512 is consistent with the feed structure. TGA tests have indicated that the thermal stability of copolymer PA56/512 increased with an increasing proportion of 512. The second heating DSC curves of PA56/512 with different composition ratios reveal the typical isodimorphism behavior of the PA56/512, the lowest melting points of which are attained with 40% 56 and 60% 512 (sample #6). The non-isothermal crystallization kinetics analysis of PA56/512, undertaken using Mo's method and Kissinger method, indicate that the crystallization ability of copolymer PA56/512 follows a similar trend to its melting temperature, with first decreasing and then increasing. The crystallization ability of pure PA512 was higher than that of PA56, which can be attributed to the high mobility and fast conformational transition of long polymethylene segments in the 512 component.

Author Contributions: Conceptualization, X.L. and L.C. (Lingna Cui); methodology, S.F. and L.Y.; validation, T.H., Y.L. and Y.W.; Formal analysis, D.G.; Investigation, D.G.; Writing—Original Draft preparation, D.G.; Writing—Review and Editing, D.G. and L.Y.; Supervision, L.C. (Lily Chen), Z.M.; Project administration, S.F.; Funding acquisition, Y.L. All authors have read and agreed to the published version of the manuscript.

Funding: This research was supported by the National Natural Science Foundation of China (No.11872179) and the Natural Science Foundation of Hunan Province (No. 2018JJ4072). The APC was funded by National Natural Science Foundation of China (No.11872179).

Institutional Review Board Statement: Not applicable.

Data Availability Statement: The following are available online in section "MDPI Research Data Policies" at https://www.mdpi.com/ethics (accessed on 10 May 2023).

Acknowledgments: This work was supported by the National Natural Science Foundation of China (No.11872179) and the Natural Science Foundation of Hunan Province (No. 2018JJ4072).

Conflicts of Interest: The authors declare no conflict of interest.

References

1. Marchildon, K. Polyamides—Still Strong after Seventy Years. *Macromol. React. Eng.* **2011**, *5*, 22–54. [CrossRef]
2. Chang, H.-H.; Beltsios, K.; Lin, D.-J.; Cheng, L.-P. Formation of polyamide 12 membranes via thermal-nonsolvent induced phase separation. *J. Appl. Polym. Sci.* **2013**, *130*, 14–24. [CrossRef]
3. Lau, W.; Gray, S.; Matsuura, T.; Emadzadeh, D.; Chen, J.P.; Ismail, A. A review on polyamide thin film nanocomposite (TFN) membranes: History, applications, challenges and approaches. *Water Res.* **2015**, *80*, 306–324. [CrossRef] [PubMed]

4. Jang, J.H.; Ahn, S.H. Numerical and experimental analysis of heat sealing of multi-layered laminate films used in lithium polymer battery packaging. *J. Plast. Film Sheeting* **2016**, *33*, 142–167. [CrossRef]
5. Lin, X.T.; Wu, Y.C.; Cui, L.N.; Chen, X.; Fan, S.H.; Liu, Y.J.; Liu, X.C.; Zheng, W. Crystal structure, morphology, and mechanical properties of biaxially oriented polyamide-6 films toughened with poly(ether block amide). *Polym. Adv. Technol.* **2022**, *33*, 137–145. [CrossRef]
6. Froidevaux, V.; Negrell, C.; Caillol, S.; Pascault, J.P.; Boutevin, B. Biobased Amines: From Synthesis to Polymers; Present and Future. *Chem. Rev.* **2016**, *116*, 14181–14224. [CrossRef]
7. Xue, Y.; Zhao, Y.; Ji, X.; Yao, J.; Busk, P.K.; Lange, L.; Huang, Y.; Zhang, S. Advances in bio-nylon 5X: Discovery of new lysine decarboxylases for the high-level production of cadaverine. *Green Chem.* **2020**, *22*, 8656–8668. [CrossRef]
8. Winnacker, M.; Rieger, B. Biobased Polyamides: Recent Advances in Basic and Applied Research. *Macromol. Rapid Commun.* **2016**, *37*, 1391–1413. [CrossRef]
9. Zhang, S.; Ma, J. Study on the unsaturated hydrogen bond behavior of bio-based polyamide 56. *E-Polymers* **2019**, *19*, 23–31. [CrossRef]
10. Huang, W.; Zhai, J.; Zhang, C.; Hu, X.; Zhu, N.; Chen, K.; Guo, K. 100% Bio-Based Polyamide with Temperature/Ultrasound Dually Triggered Reversible Cross-Linking. *Ind. Eng. Chem. Res.* **2020**, *59*, 13588–13594. [CrossRef]
11. Zou, G.; Wang, P.; Feng, W.; Ren, Z.; Ji, J. Bio-based transparent polyamide 10T/10I/1012 with high performance. *J. Appl. Polym. Sci.* **2019**, *136*, 47305. [CrossRef]
12. Lee, M.O.; Kim, J.H.; Park, J.; Kim, S.Y. Bio-based poly(pentamethylene sebacamide) by solid-state polymerization from bio-based monomers. *Green Chem.* **2021**, *23*, 6469–6476. [CrossRef]
13. Kim, H.T.; Baritugo, K.-A.; Oh, Y.H.; Hyun, S.M.; Khang, T.U.; Kang, K.H.; Jung, S.H.; Song, B.K.; Park, K.; Kim, I.-K.; et al. Metabolic Engineering of *Corynebacterium glutamicum* for the High-Level Production of Cadaverine That Can Be Used for the Synthesis of Biopolyamide 510. *ACS Sustain. Chem. Eng.* **2018**, *6*, 5296–5305. [CrossRef]
14. Gao, A.; Zhang, H.; Sun, G.; Xie, K.; Hou, A. Light-induced antibacterial and UV-protective properties of polyamide 56 biomaterial modified with anthraquinone and benzophenone derivatives. *Mater. Des.* **2017**, *130*, 215–222. [CrossRef]
15. Zhang, Y.; Wang, Y.; Xu, Y.; Liu, X.; Guo, W. Modification of biobased polyamide 56 to achieve ultra-toughening. *Polym. Technol. Mater.* **2021**, *60*, 1585–1604. [CrossRef]
16. Xu, Y.; Zhao, C.; Guo, Z.; Dong, W.; Liu, X.; Guo, W. EPDM-g-MAH toughened bio-based polyamide 56 to prepare thermoplastic polyamide elastomer and the performance characterization. *J. Appl. Polym. Sci.* **2022**, *139*, 52346. [CrossRef]
17. Yan, Y.; Gooneie, A.; Ye, H.; Deng, L.; Qiu, Z.; Reifler, F.A.; Hufenus, R. Morphology and Crystallization of Biobased Polyamide 56 Blended with Polyethylene Terephthalate. *Macromol. Mater. Eng.* **2018**, *303*, 1800214. [CrossRef]
18. Yang, T.; Gao, Y.; Liu, X.; Wang, X.; Yu, J.; Huang, S.; Ma, B.; He, Y. A Strategy to Achieve the Inherently Flame-retardant PA56 by Copolymerization with DDP. *J. Polym. Environ.* **2022**, *30*, 3802–3814. [CrossRef]
19. Yang, K.; Ma, L.; Chen, L.; Li, D.; Lu, D.; Tang, Z. Effect of Copolymerization Sequence Structure on Non-isothermal Crystallization Kinetics of Bio-based Copolymer PA 5T/56. *Eng. Plast. Appl.* **2022**, *50*, 129–135.
20. Ai, T.; Zou, G.; Feng, W.; Ren, Z.; Li, F.; Wang, P.; Lu, B.; Ji, J. Synthesis and properties of biobased copolyamides based on polyamide 10T and polyamide 56 through one-pot polymerization. *New J. Chem.* **2021**, *45*, 14677–14686. [CrossRef]
21. Wang, Y.; Kang, H.L.; Wang, R.; Liu, R.G.; Hao, X.M. Crystallization of polyamide 56/polyamide 66 blends: Non-isothermal crystallization kinetics. *J. Appl. Polym. Sci.* **2018**, *135*, 46409. [CrossRef]
22. Villaseñor, P.; Franco, L.; Subirana, J.A.; Puiggalí, J. On the crystal structure of odd–even nylons: Polymorphism of nylon 5,10. *J. Polym. Sci. Part B Polym. Phys.* **1999**, *37*, 2383–2395. [CrossRef]
23. Cui, X.; Li, W.; Yan, D. A study of the crystalline transitions of polyamides X 18. *Polym. Int.* **2004**, *53*, 2031–2037. [CrossRef]
24. Zong, R.; Hu, Y.; Liu, N.; Li, S.; Liao, G. Investigation of thermal degradation and flammability of polyamide-6 and polyamide-6 nanocomposites. *J. Appl. Polym. Sci.* **2007**, *104*, 2297–2303. [CrossRef]
25. Gonçalves, E.S.; Poulsen, L.; Ogilby, P.R. Mechanism of the temperature-dependent degradation of polyamide 66 films exposed to water. *Polym. Degrad. Stab.* **2007**, *92*, 1977–1985. [CrossRef]
26. Yang, K.; Liu, Y.; Zheng, Z.; Lu, G.; Tang, Z.; Chen, X. Synthesis and thermal degradation mechanism of a semi-aromatic copolyamide from renewable sources. *Polym. Degrad. Stab.* **2022**, *203*, 110089. [CrossRef]
27. Li, X.; He, Y.; Dong, X.; Ren, X.; Gao, H.; Hu, W. Effects of hydrogen-bonding density on polyamide crystallization kinetics. *Polymer* **2020**, *189*, 122165. [CrossRef]
28. Xia, J.F.; Xu, S.S.; Zheng, Y.; Zhou, J.; Yu, C.T.; Shan, G.R.; Bao, Y.Z.; Pan, P.J. Isodimorphic Crystallization and Tunable gamma-alpha Phase Transition in Aliphatic Copolyamides: Critical Roles of Comonomer Defects and Conformational Evolution. *Macromolecules* **2022**, *55*, 6090–6101. [CrossRef]
29. Yuan, W.; Zhou, J.; Liu, K.; Li, X.; Xu, W.; Song, H.; Shan, G.; Bao, Y.; Zhao, Q.; Pan, P. Sequence-Rearranged Cocrystalline Polymer Network with Shape Reconfigurability and Tunable Switching Temperature. *ACS Macro Lett.* **2020**, *9*, 588–594. [CrossRef]
30. Yu, Y.; Wei, Z.; Zhou, C.; Zheng, L.; Leng, X.; Li, Y. Miscibility and competition of cocrystallization behavior of poly(hexamethylene dicarboxylate)s aliphatic copolyesters: Effect of chain length of aliphatic diacids. *Eur. Polym. J.* **2017**, *92*, 71–85. [CrossRef]
31. Li, J.; Fang, Z.; Tong, L.; Gu, A.; Liu, F. Effect of multi-walled carbon nanotubes on non-isothermal crystallization kinetics of polyamide 6. *Eur. Polym. J.* **2006**, *42*, 3230–3235. [CrossRef]

32. Liu, Y.; Yang, G.S. Non-isothermal crystallization kinetics of polyamide-6/graphite oxide nanocomposites. *Thermochim. Acta* **2010**, *500*, 13–20. [CrossRef]
33. Sun, X.; Mai, K.; Zhang, C.; Cao, M.; Zhang, Y.; Zhang, X. Nonisothermal crystallization kinetics of bio-based semi-aromatic polyamides. *J. Therm. Anal. Calorim.* **2017**, *130*, 1021–1030. [CrossRef]
34. Raimo, M. "Kinematic" analysis of growth and coalescence of spherulites for predictions on spherulitic morphology and on the crystallization mechanism. *Prog. Polym. Sci.* **2007**, *32*, 597–622. [CrossRef]
35. Di Lorenzo, M.; Silvestre, C. Non-isothermal crystallization of polymers. *Prog. Polym. Sci.* **1999**, *24*, 917–950. [CrossRef]
36. Liu, T.; Mo, Z.; Wang, S.; Zhang, H. Nonisothermal melt and cold crystallization kinetics of poly(aryl ether ether ketone ketone). *Polym. Eng. Sci.* **1997**, *37*, 568–575. [CrossRef]
37. De Melo, C.C.N.; Beatrice, C.A.G.; Pessan, L.A.; de Oliveira, A.D.; Machado, F.M. Analysis of nonisothermal crystallization kinetics of graphene oxide—Reinforced polyamide 6 nanocomposites. *Thermochim. Acta* **2018**, *667*, 111–121. [CrossRef]
38. Zhang, Q.; Zhang, Z.; Zhang, H.; Mo, Z. Isothermal and nonisothermal crystallization kinetics of nylon-46. *J. Polym. Sci. Part B Polym. Phys.* **2002**, *40*, 1784–1793. [CrossRef]
39. Cui, X.; Qing, S.; Yan, D. Isothermal and nonisothermal crystallization kinetics of novel odd–odd polyamide 9 11. *Eur. Polym. J.* **2005**, *41*, 3060–3068. [CrossRef]
40. Liu, M.; Zhao, Q.; Wang, Y.; Zhang, C.; Mo, Z.; Cao, S. Melting behaviors, isothermal and non-isothermal crystallization kinetics of nylon 1212. *Polymer* **2003**, *44*, 2537–2545. [CrossRef]
41. Puiggalí, J.; Franco, L.; Alemán, C.; Subirana, J.A. Crystal Structures of Nylon 5,6. A Model with Two Hydrogen Bond Directions for Nylons Derived from Odd Diamines. *Macromolecules* **1998**, *31*, 8540–8548. [CrossRef]
42. Deshmukh, Y.S.; Wilsens, C.H.R.M.; Verhoef, R.; Hansen, M.R.; Dudenko, D.; Graf, R.; Klop, E.A.; Rastogi, S. Conformational and Structural Changes with Increasing Methylene Segment Length in Aromatic–Aliphatic Polyamides. *Macromolecules* **2016**, *49*, 950–962. [CrossRef]

polymers

MDPI

Article

Determination of Sustainable Critical Flux through a Long-Term Membrane Resistance Model

Rongle Xu [1,2,*], Yaobo Fan [3], Min Yang [3] and Jinqiu Song [2]

1 Scientific Research Academy of Guangxi Environmental Protection, Nanning 530022, China
2 School of Resources, Environment and Materials, Guangxi University, Nanning 530004, China; 17745056451@163.com
3 Research Center for Eco-Environmental Sciences, Chinese Academy of Sciences, Beijing 100085, China; ybfan@rcees.ac.cn (Y.F.); yangmsunb@126.com (M.Y.)
* Correspondence: xurongle@gxhky.org

Abstract: A long-term membrane resistance model (LMR) was established to determine the sustainable critical flux, which developed and simulated polymer film fouling successfully in a lab-scale membrane bioreactor (MBR) in this study. The total polymer film fouling resistance in the model was decomposed into the individual components of pore fouling resistance, sludge cake accumulation and cake layer compression resistance. The model effectively simulated the fouling phenomenon in the MBR at different fluxes. Considering the influence of temperature, the model was calibrated by temperature coefficient τ, and a good result was achieved to simulate the polymer film fouling at 25 and 15 °C. The relationship between flux and operation time was simulated and discussed through the model. The results indicated that there was an exponential correlation between flux and operation time, and the exponential curve could be divided into two parts. By fitting the two parts to two straight lines, respectively, the intersection of the two straight lines was regarded as the sustainable critical flux value. The sustainable critical flux obtained in this study was just 67% of the critical flux. The model in this study was proven to be in good agreement with the measurements under different fluxes and different temperatures. In addition, the sustainable critical flux was first proposed and calculated in this study, and it was shown that the model could be used to predict the sustainable operation time and sustainable critical flux, which provide more practical information for designing MBRs. This study is applicable to polymer films used in a wide variety of applications, and it is helpful for maintaining the long-term stable operation of polymer film modules and improving the efficiency of polymer film modules.

Keywords: sustainable critical flux; membrane resistance model; polymer film fouling

Citation: Xu, R.; Fan, Y.; Yang, M.; Song, J. Determination of Sustainable Critical Flux through a Long-Term Membrane Resistance Model. *Polymers* **2023**, *15*, 2319. https://doi.org/10.3390/polym15102319

Academic Editor: Alfredo Cassano

Received: 21 March 2023
Revised: 26 April 2023
Accepted: 27 April 2023
Published: 16 May 2023

1. Introduction

Membrane bioreactors (MBRs) offer a number of advantages over the conventional activated sludge process, such as excellent quality effluent, a more compact treatment facility, a more concentrated biomass and a reduced sludge yield [1]. In the application of wastewater treatment and water reclamation, MBRs have progressively gained acceptance and popularity in China [2]. However, polymer film fouling remains a big challenge for widespread applications of MBRs, which can reduce polymer film flux at a given transmembrane pressure (TMP), or conversely, increase the TMP at a given flux, and finally, increase the energy consumption and costs of the wastewater treatment and water reclamation.

Membrane cleaning is important to reduce membrane contamination and improve membrane performance. Gul et al. [3] reviewed the fouling and cleaning process in microfiltration membranes. Polymer film fouling in a submerged MBR can be attributed to both polymer film pore clogging and sludge cake deposition on the polymer film surface [4]. Numerous studies have directly elucidated the fouling behaviors, and the effects of the key operation conditions and sludge characteristics on polymer film fouling. Recent experimental results indicate that polymer film resistance is mainly affected by aeration turbulence and sludge characterization [5–8]. Mixed liquor suspended solids (MLSS) and colloids are the main contributors to polymer film resistance [9]. Polymer film pore clogging could be caused by the adsorption of dissolved or colloidal matters, and soluble microbial products (SMP) have been proved to be the dominant pollutant causing pore blocking during long-term operation [10,11]. MLSS were found to be the main contributor to the cake layer [12]. Aeration has a positive effect on cake layer removal, and a higher air flow rate could promote the back transport of deposited materials from the polymer film surface by turbulent shear [13] but with higher energy consumption.

Operation flux is an important parameter in MBR designing and operation, and it is a direct factor in polymer film fouling. Hung et al. [14] reviewed recent studies on membrane compaction, wetting and fouling, demonstrating that an ultra-low-pressure membrane (ULPM) is effective for long-term filtration at stable fluxes. It has been suggested that submerged MBRs should be operated at a flux below the "critical flux", the so-called subcritical flux, to maintain a sustainable permeability and to mitigate polymer film pollution [15–17]. The critical flux or subcritical flux is a variable with a relatively wide range, and there is no clear criterion for accurate subcritical flux determination. For an MBR system, even operating at subcritical fluxes there are significant differences in the polymer film fouling rates and the community structures of the sludge layer that adhere to the polymer film surface [18–20]. What is more, based on the sustainable critical flux ($flux_{SC}$), the necessary polymer film area can be determined. In the MBR, too much polymer film, which is determined by a lower flux, increases the investment, but not enough polymer film, which is determined by a higher flux, reduces the output at the designed scale and leads to difficulty maintaining a stable performance on the high flux compared with the $flux_{SC}$. To maintain an MBR's long-term stable operation, it is important to predict the behaviors of polymer film fouling and cake formation during its long-term operation, and it is necessary to limit the scope of subcritical fluxes, and thus find a sustainable critical flux.

A polymer film filtration model is able to accurately express the polymer film fouling characterization, which allows for determination of the independent fouling characteristics and a better fundamental understanding of the controlling mechanisms. Model-based analysis is an essential tool for polymer film fouling control and the long-term stable operation of an MBR. The polymer film filtration or polymer film fouling models can be classified into two groups, the mechanistic models and the mathematic models. The mechanistic models are based on physical filtration laws, and resistances-in-series is the most common approach [21,22]. The mathematic models based on statistical analysis are used to reveal the relationship among aeration, flux and sludge characterization [23,24]. However, some of the models are restricted in relatively short-term periods, and others cannot quantify the contributions of polymer film foulants to overall polymer film fouling. Moreover, most of these models can neither predict the trend of long-term polymer film fouling nor guide the design and the operation of an actual MBR.

The objective of this study is to predict the sustainable critical flux of A^2/O-MBR by establishing a long-term membrane resistance model. The sustainable critical flux could be used as operating parameter, which could extend membrane service life and control membrane fouling effectively.

2. Theories and Models

In this study, the resistance R was calculated according to Darcy's law,

$$R = TMP/\mu J \quad (1)$$

where R is the resistance, m^{-1}, and TMP is the pressure, kPa. μ is the viscosity of the sludge suspension, mPa•s. J is the flux, $m^3/(m^2•d)$.

The long-term membrane resistance model established in this study is based on a two-stages hypothesis. At the first stage, the specific flux declined rapidly and the operation TMP, TMP_a, grew slowly along with the operation time; the polymer film pore clogging caused by SMP and the cake layer formation were the main polymer film fouling processes. At the second stage, as TMP_a reached a certain value (the critical TMP, TMP_c), the specific flux declined slowly and TMP_a rose rapidly; cake layer compression was the main polymer film fouling process.

2.1. Components of Filtration Resistance

The total resistance R comprises three resistance components, including the intrinsic resistance of the polymer film, R_m, the pore fouling resistance caused by solute deposition inside the polymer film pores, R_p and the resistance of the sludge cake layer, R_c, which is formed by the resistance of sludge cake accumulation R_{ac} and the resistance of cake layer compression R_{comp}. R, R_m, R_p, R_{ac} and R_{comp}, m^{-1}.

When $TMP_a < TMP_c$,

$$R = R_m + R_p + R_{ac} \tag{2}$$

and when $TMP_a \geq TMP_c$,

$$R = R_m + R_p + R_{comp} \tag{3}$$

2.2. Polymer Film Pore Fouling Resistance (R_p)

The pore fouling resistance is expected to increase proportionally to the volume of water production during filtration [25]; that is,

$$R_p = r_p J C t \tag{4}$$

where r_p is the specific pore fouling resistance, m/kg. C is the concentration of polymer film pore foulants, kg/m^3. J is the operation flux, $m^3/(m^2 \bullet h)$. t is the operation time, h.

Many studies have shown that SMP is the main foulant of pore clogging [11,26]; thus, in this paper, we chose the SMP to calculate the R_p.

$$R_p = r_{SMP} J C_{SMP} t\theta \tag{5}$$

where r_{SMP} is the specific pore fouling resistance, m/kg. C_{SMP} is the concentration of polymer film pore foulants, kg/m^3. θ is the coefficient of polymer film pore clogging caused by SMP.

In this study, through the filtration experiments with different sludge components, SMP was also proved to be the main factor contributing to polymer film clogging, of which the contribution rate could be more than 50%, and the value of θ was 1~2, which was influenced by the suction drag force.

2.3. Sludge Cake Accumulation Resistance (R_{ac})

The resistance of the sludge cake layer, R_{ac}, was calculated by Equation (5),

$$R_{ac} = r_c \times M_C \tag{6}$$

where r_c is the specific filtration resistance of the sludge cake layer, m/kg. M_C is the amount of MLSS accumulating on the polymer film surface, kg/m^2. M_C is a variable changed along with the operation time, and affected by the aeration. The calculation of M_C is discussed with the sludge particle being deposited on the polymer film surface and detaching from the polymer film surface.

1. Probability of suspended solids being deposited on the polymer film surface (E).

Two opposite forces regulate the probability (E) of the sludge particle being deposited on the polymer film surface, the drag force and the lifting force, of which the drag force caused by suction leads to the attachment, and the lifting force caused by the turbulent flow prevents the sludge particles from attaching to the polymer film surface (Figure 1a). E can be estimated [27] as

$$E = \frac{24J}{24J + K_1 G} \quad (7)$$

where K_1 is a constant that is related to the lifting force of the sludge particle and particle size, m. $K_1 = 4 \times 10^{-6}$ m [27], and G is the shear intensity on the polymer film surface, s^{-1}, which can be determined by Equation (8).

$$G = \left(\frac{\rho g q}{\mu}\right)^{1/2} \quad (8)$$

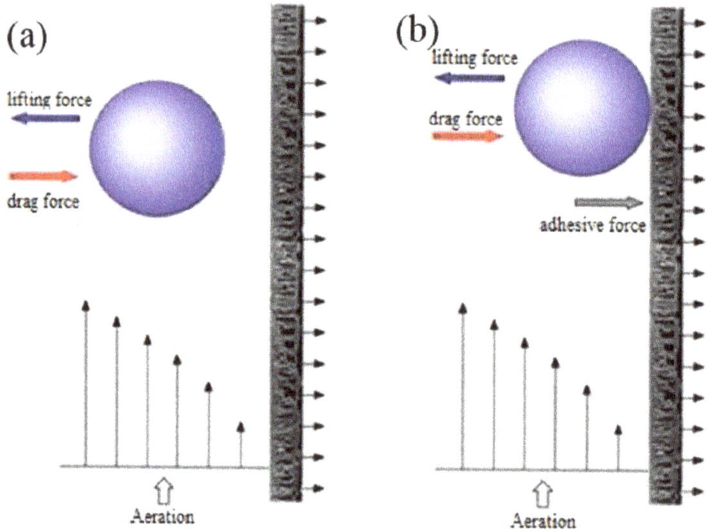

Figure 1. Force diagram of sludge particle. (**a**) Sludge particle being deposited on the polymer film surface, (**b**) sludge particle attached to the polymer film surface.

g is the gravitational constant, 9.81 m/s^2. q is the aeration intensity, $L/(m^2 \bullet s)$. ρ is the density of biomass particles, kg/m^3 and μ is the viscosity, $Pa \bullet s$.

It is noted that the sludge particles can be affected by the drag force, and the depositing of the sludge particles on the polymer film surface is only easier when the particle size is less than 100 μm [28]. Thus, a coefficient ε, the percent of the sludge practice size less than 100 μm, was introduced into the model.

Above all, the rate of biomass attachment onto the polymer film surface can be written as

$$\frac{dM_c}{dt} = ECJ\varepsilon \quad (9)$$

where C is the sludge concentration, kg/m^3. E is the probability of the suspended solids being deposited on the polymer film surface (Equation (7)), J is the operation flux, $m^3/(m^2 \bullet h)$, and ε is the percentage of the sludge practice size less than 100 μm.

2. Probability of the suspended solids detaching from the polymer film surface (K).

When a sludge particle attaches on the polymer film surface, there are three forces regulating the probability, the drag force, the lifting force and the adhesive force caused by the stickiness between the biomass particles and the polymer film surface or between the biomass particles (Figure 1b).

However, there is a boundary layer adjacent to the polymer film surface where the velocity is quite low, with a very weak lifting force for the biomass particles to be flushed off from the polymer film surface. Hence, a coefficient γ was introduced into the model to express the ratio of the velocity in the boundary layer (V_{bl}) to that of the bulk flow (V_{bf}). A stickiness value α was introduced, and $1-\alpha$ indicates the actual scouring coefficient after overcoming the adhesion force between the particles and the polymer film surface. Three forces regulate the probability (K) of the sludge particle detaching from the polymer film surface, and K can be estimated by

$$K = \frac{\gamma K_1 G}{\gamma K_1 G + 24J}(1-\alpha) \tag{10}$$

where α is the stickiness between the sludge particle and the polymer film surface, $\alpha = 0.1$ [29]. γ is the ratio of V_{bl} to V_{bf}.

The rate of biomass detachment from the polymer film surface can be determined by

$$\frac{dM_f}{dt} = KCJ\varepsilon \tag{11}$$

where C is the sludge concentration, kg/m^3, K is the probability of SS detaching from the polymer film surface, J is the operation flux, $m^3/(m^2 \bullet h)$, and ε is the percentage of the sludge practice size less than 100 μm.

The M_C in Equation (5) can be determined by Equations (8)–(10) and by integrating $d(M_c - M_f)$, and finally, the sludge cake accumulation resistance. Equation (5) can be established as

$$R_c = r_c(Et_1 - Kt_2)CJ\varepsilon \tag{12}$$

where t_1 is the time of polymer film suction and t_2 is the aeration time.

2.4. Cake Layer Compression Resistance (R_{comp})

The sludge cake is highly compressible and its properties vary with the sharp increase in the applied TMP. Cake layer compression may lead to an increase in the sludge specific resistance, r_{comp}, which can be described by the empirical Equation (13) [30],

$$r_{comp} = r_c\left(1 + \frac{TMP_a}{TMP_c}\right)^n \tag{13}$$

where TMP_c, the critical TMP, is the TMP at which r_c is double its initial value, and the value of TMP_c varies with the sludge type, sludge conditioning, floc size, etc. [31]. TMP_a is the applied TMP. The compressibility coefficient, n, value is in the range of 0 to 1, and for activated sludge, $n = 1$ [31]. At the stage of cake layer compression, the sludge biomass is supposed to cover the polymer film surface fully. Therefore, because of the higher stickiness of sludge biomass, the value of the stickiness coefficient α is set to 0.5 [32].

Based on Equations (12) and (13), the cake layer resistance at the compression stage is expressed as Equation (14),

$$R_{comp} = r_{comp}(Et_1 - Kt_2)CJ\varepsilon \tag{14}$$

3. Experiments

3.1. Lab-Scale Setup

A submerged A^2/O-MBR was operated to characterize the polymer film fouling under various conditions (Figure 2). The experimental setup was made up of an anaerobic tank, an anoxic tank and an oxic tank with submerged polymer film modules. The working volume of the A^2/O-MBR was 58.3 L, including anaerobic tank 8.3 L, anoxic tank 16.7 L and aerobic tank 33.3 L. The anaerobic and anoxic tanks were mixed with a low-speed mixer and air was introduced via perforated pipe under the polymer film module in the oxic tank. Three pieces of flat-sheet PVDF polymer films (0.1 m^2 per polymer film element, membrane pore size 0.2 μm, SINAP membrane S&T Co., Shanghai, China) were used in the oxic tank of A^2/O-MBR.

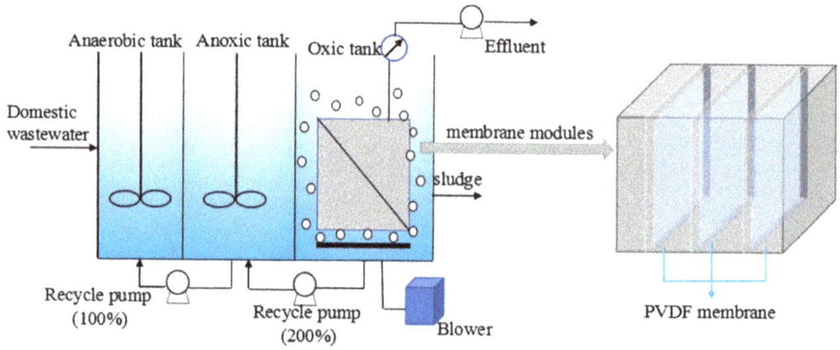

Figure 2. Schematic of lab-scale A^2/O-MBR reactors.

The influent to the MBR was domestic wastewater taken from the sewer at residential district of Research Center for Eco-Environmental Science, Chinese Academy of Science. The sludge residence time (SRT) in A^2/O-MBR was approximately 20 days or longer. The sludge in the MBR was maintained at a steady state for more than 2 months before experiments, and the fouling experiments lasted more than 2 months. The sludge concentration was about 6 g/L.

Critical flux was determined by the flux-step method [16]. Under trial condition, the critical flux was 30 LMH. Model simulation was carried out with two-stage experiments. The first was operated at different flux under average temperature 15 °C. The filtration flux was set under critical flux, and 10, 18 and 25 LMH were adopted. The other was operated at 10 LMH under average temperature 25 °C. The polymer films operated under a 6 min filtration and 1 min stop cycle; no backwashing of the polymer film was required. The aeration intensity was fixed at 0.83 L/(m^2•s).

During the test, the increase in the TMP was recorded, along with the running time and water production. The filtration flux of the water production was monitored every day. The MBR stopped when the TMP reached 30 kPa, and then the polymer film module was cleaned thoroughly by physical and chemical methods. Polymer film resistance analyses was calculated by resistance series models in Feng's study [33]. The irreversible resistance that could not be cleaned by physical and chemical methods was ignored in this study because the new polymer film was used. The physical washing removed cake layer by showering with tap water and sponge swabbing; by this step, the R_c was calculated. Chemical washing removed the foulants from polymer film pore, and the R_p was calculated by this step.

3.2. *Model Parameters' Determination*

3.2.1. R_p and r_c

According to ultrafiltration cup experiment, there is a linear relationship between filter time and volume within a certain range. Specific filtration resistance r can be described as [8]

$$\frac{t}{V} = \left(\frac{\mu r C}{2PA^2}\right)V + \frac{\mu R_m}{PA} \tag{15}$$

The slope b = $\frac{\mu r C}{2PA^2}$, r = $\frac{2PA^2 b}{\mu C}$

where μ is the permeate viscosity, mPa•s. C is the concentration of mixed liquor suspended solids, kg/m^3. A is the effective filtration area, m^2. P is the filtration pressure, Pa. t is the filtration time, s. V is the filtration volume, m^3.

3.2.2. Ratio of Boundary Velocity to Bulk Flow Velocity

A simulation of the 3D flow field between polymer film sheets by the commercial CFD code ANSYS Fluent® showed that there is non-uniformity of gas–liquid flow in channels between two polymer film elements; the flow velocity is higher in the middle of the channel and that much lower in the boundary layer at the polymer film surface [34]. According to the simulation, the ratio of boundary velocity to bulk flow velocity γ is selected at 0.02.

3.2.3. Temperature Coefficient τ

Temperature is a key factor influencing the microbial community, sludge morphology and the filtration of mixed liquid in MBR operation [35]; interactions between temperature, sludge characterization and polymer film fouling are of a complex nature [36]. Low temperature might increase mixed liquor viscosity and reduce particle size, which causes lower particle back transport velocity, and furthermore, leads to negative influence on the polymer film performance [35,37]. Considering the complex role of temperature in polymer film fouling, it is necessary to introduce a coefficient for temperature calibration. Thus, a coefficient τ was introduced in the cake resistance calculation; the Equations (12) and (13) can be changed to

$$R_c = r_c(\mathrm{E}t_1 - \mathrm{K}t_2)\mathrm{CJ}\varepsilon\tau \quad (16)$$

$$R_{comp} = r_{comp}(\mathrm{E}t_1 - \mathrm{K}t_2)\mathrm{CJ}\varepsilon\tau \quad (17)$$

Some studies suggested differences in permeability of 50% were found between summer and winter periods in the full-scale MBR [37,38]. A lab-scale study also found the differences in permeate flux and permeability between the low temperature and high temperature were from 30% to 80% [39]. Therefore, in this study, the value of τ was set as 0.3.

In this study, a lab-scale experiment was carried out to prove the effect of temperature with flux set at 10 LMH and temperature at 25 °C. The total polymer film fouling resistances under different simulations are shown in Figure 3 and Table 1. It can be seen that after calibration by coefficient τ, the model simulates the real performance very well, the average deviation between simulated and measured values is 7.18%, which means that the influence of temperature on polymer film fouling can be reflected by coefficient τ.

The parameters and coefficients for model were selected based on laboratory tests, assumptions and literature reports (Table 2). The sludge concentration, SMP concentration and μ were averages of measurements taken during the whole test. SMP was filtered (0.2 µm, cellulose acetate polymer film) and measured TOC (Vario TOC, Elementar Analysensysteme GmbH, Germany). ε, the percent of sludge practice size less than 100 µm, was tested by laser particle size analyzers (Malvern Mastersizer2000, Malvern Instruments Ltd., Malvern, UK).

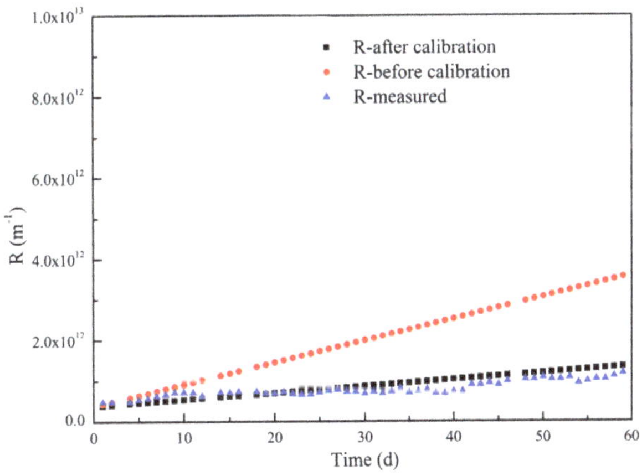

Figure 3. Temperature calibration of long-term membrane resistance model.

Table 1. Comparison of resistances before and after temperature calibration.

Measured Value (m^{-1})	Simulated Value (m^{-1})		Standard Deviation (%)	
	Before Calibration	After Calibration	Before Calibration	After Calibratration
1.18×10^{12}	2.73×10^{12}	1.32×10^{12}	101.57 ± 63.90	7.18 ± 21.86

Table 2. Parameters and coefficients used in the modelling and simulations.

Symbol	Meaning	Value	Source of Data
C	Sludge concentration, (MLVSS, g/L)	6.5	Measured
C_{SMP}	SMP concentration, (TOC, mg/L)	37.7	Measured
g	Gravitational constant (m/s^2)	9.81	Constant
J	Flux ($m^3/(m^2 \bullet d)$)	——	Operation parameter
K_1	Coefficient (m)	4×10^{-6}	Constant (from literature)
q	Aeration intensity ($L/(m^2 \bullet s)$)	0.83	Operation parameter
R_m	Polymer film intrinsic resistance (m^{-1})	3.74×10^{11}	Measured
r_{SMP}	Specific pore fouling resistance (m/kg)	9.1×10^{11}	Measured
r_c	Specific filtration resistance of cake laye (m/kg)	3.14×10^{11}	Measured
TMP_c	Critical pressure (kPa)	10	Measured
t_1	Filtration time (d)	6 min filtration and 1 min	Operation parameter
t_2	Aeration time (d)	stop cycle, $t_1 = 6/7t_2$.	Operation parameter
α	Stickiness of biomass particles or stickiness between biomass particles and polymer film surface	At sludge accumulation stage, $\alpha = 0.1$; at cake layer compression stage, $\alpha = 0.5$	Constant (from literature)
γ	The ratio of boundary velocity to flow velocity	0.02	Constant (from previous work)
ε	The percent of sludge practice size less than 100 μm	0.6	Measured
μ	Viscosity of the sludge suspension ($mPa \bullet s$)	5.5	Measured
θ	Coefficient of polymer film pore clogging caused by SMP	1.25 at 10 LMH, 1.5 at 18 and 25 LMH	Constant (from previous work)
ρ	Density of the sludge suspension (kg/m^3)	1.0×10^3	Constant

4. Results and Discussion

4.1. Model Simulation under Different Fluxes in the Lab-Scale A^2/O-MBR

TMP and flux were monitored along time under different filtration fluxes, and the polymer film resistance R was calculated by Darcy law. With consideration of the sludge's attachment to and detachment from the polymer film surface, the long-term resistance model developed in this study is capable of simulating polymer film fouling of the lab-scale A^2/O-MBR. The individual fouling components and total polymer film resistance were obtained by simulation, and the resistances under different fluxes are shown in Figure 4.

From Figure 4, we can note that the simulation results are comparable with the experimental observations on the trend of polymer film development under different operation fluxes. For the individual component of polymer film fouling resistances, the resistance caused by sludge cake accumulation (R_c) was the dominant component of the total filtration resistance R. It is noted that R_c/R was less than 30% at the initial stage, while it was over 90% at the end of the simulation, which indicates that sludge cake accumulation and compression were sustained. Pore fouling resistance R_p was also increased along with the simulation or experiment time, although the contribution of R_p to R varied from 10% at the initial stage to 3% at the end of the simulation.

The comparison of the simulated and experimentally measured resistances under different operations is shown in Table 3. At the end of the operation time, the errors of the fouling resistance R between the simulated and experimental measurements were less than 13.5% under the different fluxes, which implied the availability and the good accuracy of the long-term membrane resistance model when used in the simulation of the fouling resistances in MBR operation. The simulated values of pore fouling resistance R_p were lower than the measured values, which resulted from the constant SMP concentration and lower value of θ set in the model, while, in fact, the SMP concentration could not be an exact constant during the A^2/O-MBR's operation.

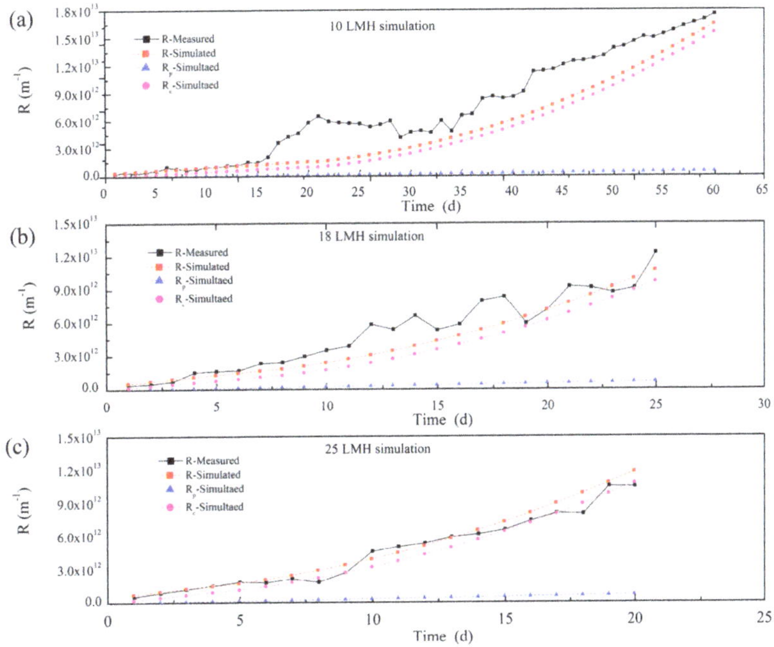

Figure 4. R simulations under different fluxes. (**a**) 10 LMH; (**b**) 18 LMH; (**c**) 25 LMH s.

Table 3. Comparison of resistances between experimental measurements and simulation under different operations.

Operational Flux	Measured Value (m^{-1})			Simulated Value (m^{-1})			Error (%)		
	R_p	R_c	R	R_p	R_c	R	R_p	R_c	R
10 LMH	5.9×10^{11}	1.58×10^{13}	1.75×10^{13}	5.30×10^{11}	1.55×10^{13}	1.65×10^{13}	−10.2	−1.9	−5.7
18 LMH	8.7×10^{11}	1.01×10^{13}	1.13×10^{13}	6.84×10^{11}	9.72×10^{12}	1.07×10^{13}	−21.4	−3.8	−5.3
25 LMH	7.3×10^{11}	9.26×10^{12}	1.04×10^{13}	6.78×10^{11}	1.08×10^{13}	1.18×10^{13}	−7.1	16.6	13.5

According to the long-term membrane resistance model, the polymer film resistances are directly influenced by t flux, TMP, sludge and SMP concentrations, and some operational are parameters listed in Table 2. A faster resistance growth could have resulted from higher flux. It was noted that the experiments with higher flux usually ended at a lower total resistance R, along with a higher pore fouling resistance R_p. Conversely, under a lower flux, the experiments always finished at a higher cake layer resistance R_c accompanied with a lower pore fouling resistance R_p. It means that the flux affected the pore fouling, and more foulants entered into the polymer film pore under the suction drag force. The cake layer resistance was influenced by the flux, sludge concentration and aeration intensity. In this study, we found that flux is an important factor for cake layer resistance. Though R_c was higher under the operation of a lower flux and lower TMP, the reactor could still run for a long time, which may be related to the difference in the bio-cake structure caused by cake accumulation and compression [19].

4.2. The Determination of Sustainable Critical Flux

The main objective of this paper was to predict the sustainable critical flux of MBR by using the long-term membrane resistance model. To search for the sustainable critical flux of MBR, the total polymer film resistance, R, with operation time were simulated by using the Equations (16) and (17). Under the conditions of different fluxes from the subcritical flux (10 LMH) to the supercritical flux (35 LMH), temperatures were set at 15 and 25 °C, and the sludge concentration was set at 8000 mg/L, with other parameters as shown in Table 2.

The specific flux actually decreased with the resistance growth and increase in TMP; thus, it can be used to measure the degree of polymer film fouling. One study has suggested that the polymer film should be cleaned while the specific flux is below 0.5 LMH/kPa [40]. According to the Darcy law, the relationship between polymer film resistance and the specific flux can be expressed as Equation (18)

$$J_s = u\frac{1}{\mu}\cdot\frac{1}{R} \tag{18}$$

where J_s is the specific flux, LMH/kPa. R is the total polymer film resistance, m^{-1}. u is the unit conversion factor.

According to the above suggestion, each simulation of MBR in this study was stopped as the specific flux reached 0.5 LMH/kPa, and the actual operation time was taken as the sustainable operation time for the given operation conditions. The total resistance curves of the modelling results are given in Figure 5, and the relationship between the flux and sustainable operation time is shown in Figure 6.

The total resistance under different fluxes was simulated by the long-term membrane resistance model. Simulated conditions were varied with different operational fluxes from the subcritical flux to the supercritical flux, 10 to 35 LMH, respectively, a sludge concentration of 8000 mg/L, and other simulated parameters as shown in Table 1. Then the sustainable operation time was predicted and is shown in Figure 6.

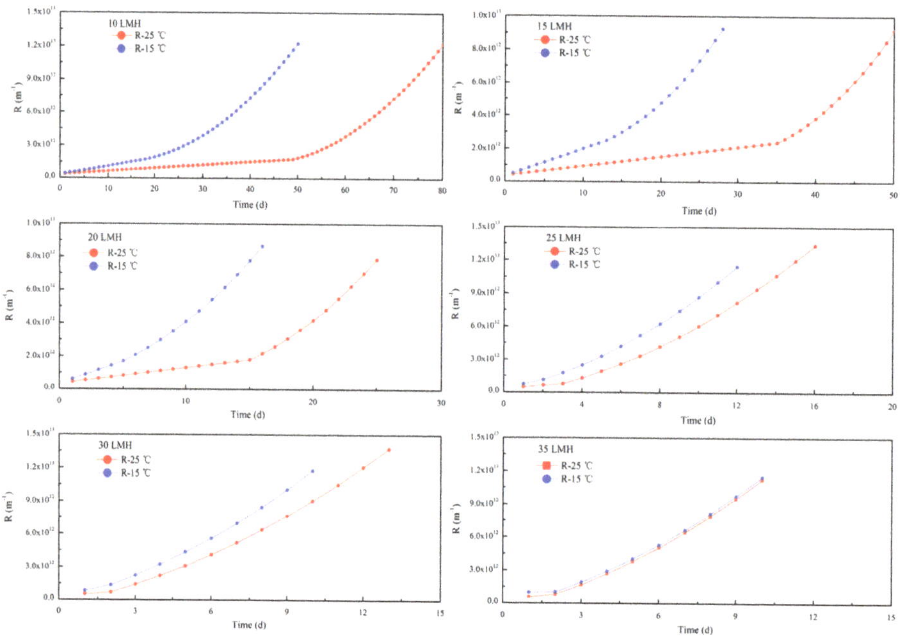

Figure 5. Total polymer film resistances under different fluxes and different temperatures.

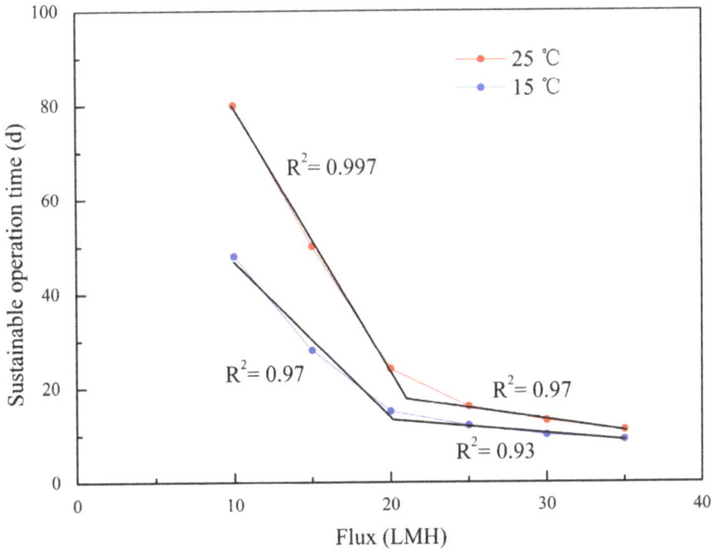

Figure 6. Relationship between flux and sustainable operation time.

From Figures 5 and 6, it can be seen that when the fluxes are in the range of 10~20 LMH, the differences in the sustainable operation times at the temperatures between 25 and 15 °C are significant. However, with the increase in flux, the sustainable operation times become shorter and their differences become smaller, and in particular, as the fluxes rise up to 35 LMH, the sustainable operation times reduce to about 10 days without a great difference at different temperatures. The result shows that temperature had an obvious effect on the total polymer film resistance at a lower flux than that at a higher flux, and the sustainable operation time under 25 °C was twice that under 15 °C. However, the differences between 25 and 15 °C were gradually narrowed with an increase in the flux. This indicates that the operation's flux may have more influence on polymer film fouling than the temperature.

The relationship between the flux and the sustainable operation time was analyzed based on the modelled data above (Figure 6). Some studies showed that the relationship between the flux and operation period was a linear correlation under subcritical fluxes in a flat-sheet MBR [24], or in a hollow fiber MBR [41]. However, different to the result of the linear correlation suggested by Wang and Guglielmi, it was approximately regarded as an exponential correlation in this study. The equations obtained under 25 and 15 °C can be expressed as Equations (19) and (20) with correlation coefficients R^2 0.985 and 0.993, respectively.

$$T_{25\ ^\circ C} = 265.85 \times e^{\left(-\frac{J}{7.86}\right)} + 6.51 \quad (19)$$

$$T_{15\ ^\circ C} = 185.94 \times e^{\left(-\frac{J}{6.56}\right)} + 7.83 \quad (20)$$

where $T_{15\ ^\circ C}$ *and* $T_{25\ ^\circ C}$ are the sustainable operation times at the temperatures of 25 °C and 15 °C, respectively, d, and J is the polymer film flux, LMH.

In Figure 6, the exponential curve can be divided into two parts, which can be fitted for two straight lines, respectively. The impact of flux on the sustainable operation time could be associated with the slope of the straight line. It is interesting to note that there is an intersection of two straight lines. When the operation flux is above this intersection value, the sustainable operation time will be shortened dramatically, and if the operation flux is below this value, the sustainable operation time will be increased significantly. The intersection of the two straight lines could be regarded as the sustainable critical flux value. Compared to the critical flux, the sustainable critical flux reflected the real relationship between the flux and the operation time during a long-term run, and narrowed the range of the subcritical flux.

The sustainable critical flux can be obtained as 20 LMH by the method described above. A frequent polymer film cleaning procedure must be conducted if the operation flux is higher than 20 LMH. It is interesting that the sustainable critical flux can be taken as the same value of 20 LMH at 25 and 15 °C, but the difference is the sustainable operation time is 25 d at 25 °C, while it is only 15 d at 15 °C. That implies that when the sustainable critical flux is determined, the temperature will affect the sustainable operation time, which is shorter at a lower temperature and longer at a higher temperature.

It is also worth noting that the sustainable critical flux obtained is at 67% of the critical flux detected in this study, and it is comparable with the "proper flux" at 56% of the critical flux obtained in the study of Wang et al. [24]. The result indicates that the critical flux was too large to be used as the sustainable operation flux. Compared with the well-known definition of critical flux, the sustainable critical flux is much more available and more practical for the design and operation of an MBR than the critical flux or subcritical flux. This is because the sustainable critical flux not only reflects the real relationship between the flux and operation time during the long-term operation of an MBR very well, but also the temperature's influence on it.

4.3. Model Application in a Large-Scale A^2/O-MBR Municipal Wastewater Treatment Plant

The model's simulation and application was conducted in a large-scale A^2/O-MBR WWTP in Beijing. The scale of the WWTP is 15,000 m^3/d, and hollow fiber polymer film modules are used with a total polymer film area 396,000 m^2. The polymer film operation parameters include alternate aeration with low and high intensities, and chemically enhanced backflush (CEB). The average intensity of high aeration is about 180 Nm^3/m^2h, while the low aeration intensity is about 80 Nm^3/m^2h. The ratio of alternate aeration time between low and high intensity is 4:1. CEB frequency is once a week with a NaClO concentration of 1500 mg/L. The average operation flux is 20 LMH. R_m of the hollow fiber polymer film is 6×10^{11}. The average concentration of SMP is about 20 mg/L. According to the operation's condition, the parameters of the polymer film resistance model were calculated and are shown in Table 4.

Table 4. Parameters of polymer film resistance model.

Parameters	r_c (m/kg)	r_p (m/kg)	γ	E	K	TMP_c (kPa)
High aeration intensity	2.97×10^{11}	8.30×10^{11}	0.01	0.125	0.059	14~15
Low aeration intensity				0.087	0.086	

CEB has positive effects on polymer film fouling control [42]. However, the mechanism of CEB is not clear. To investigate this, the effect of periodic CEB was assumed as uniform for polymer film resistance control. Two parameters, b1 and b2, were introduced into the model to represent the percent of R_p and R_c residue after CEB, respectively. Thus, the Equations (4), (16) and (17) can be expressed as Equations (21)–(23), respectively.

$$R_p = r_{SMP} J C_{SMP} t\theta b1 \tag{21}$$

$$R_c = r_c(\mathrm{E}t_1 - \mathrm{K}t_2)\mathrm{CJ}\varepsilon\tau \tag{22}$$

$$R_{comp} = r_{comp}(\mathrm{E}t_1 - \mathrm{K}t_2)\mathrm{CJ}\varepsilon\tau \mathrm{b2} \tag{23}$$

The data of three polymer film modules were used to determine b1 and b2, and b1 = 0.05, b2 = 0.95 were chosen as the proper parameters for simulation (the details of b1 and b2 determination are shown in the Supplementary Materials). The b1 value means that the CEB has a significant effect on polymer film pore clogging control, and almost 95% of

R_p can be removed by CEB, while for b2, the CEB has little effect on cake layer control, as only 5% of R_c can be removed.

The total resistances under different fluxes of the WWTP were simulated by the long-term membrane resistance model, and the relationship between the flux and the sustainable operation time was analyzed based on the simulation data as in the methods above. The simulated conditions varied the operational flux from a subcritical flux to a supercritical flux, 20 to 40 LMH, respectively, with a sludge concentration of 6000 mg/L, and other simulated parameters as shown in Table 3. Then the sustainable operation time was predicted and is shown in Figure 7.

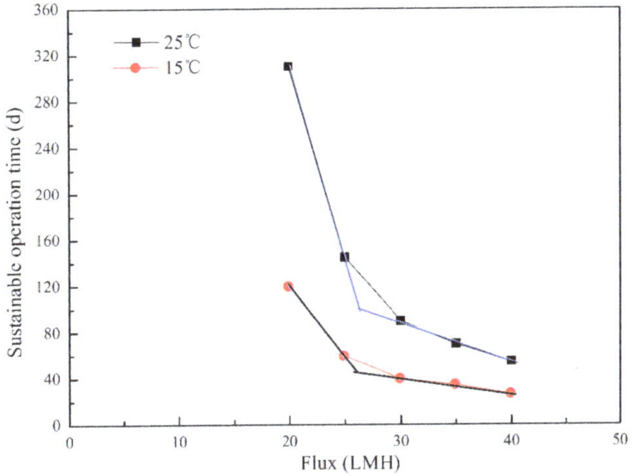

Figure 7. Relationship between flux and sustainable operation time of WWTP.

Calculating the sustainable critical flux using the data from Figure 7, it can be found that the sustainable critical flux was 26 LMH, and the polymer film module can operate stably for more than 110 days at 25 °C, while it can only operate for 48 days at 15 °C. To achieve the same operation time as that at 25 °C, the sustainable critical flux was 20 LMH at 15 °C, and this is only 75% of the sustainable critical flux at 25 °C. As we know, two factors influence the polymer film area, the operation flux and the capacity of the WWTP. In northern China, winter is the dry season with a low temperature, and should be especially considered for the determination of the polymer film area.

5. Conclusions

A long-term membrane resistance model was developed to simulate polymer film fouling of an MBR successfully. Drag force by suction, lifting force by aeration and adhesive force by stickiness were considered in the model development. The total polymer film fouling resistance in the model was decomposed into the individual components of pore fouling resistance, sludge cake accumulation and cake layer compression resistance. The model can effectively simulate the fouling phenomenon in an MBR at different fluxes. Considering the influence of temperature, the model was calibrated by temperature coefficient τ, and a good result was obtained. Additionally, the effect of flux on the polymer film operation time can be determined, and the results indicated that there was an exponential correlation between flux and operation time. The exponential curve was divided into two parts, and fitted to two straight lines, respectively, and the intersection of the two straight lines was regarded as the sustainable critical flux. The sustainable critical flux obtained in this study was just 67% of the critical flux in this study. Compared with selected operation fluxes by critical flux testing, the sustainable critical flux provides more information that is practical for the design of MBRs. In addition, the proposed model was used to simulate a full-scale WWTP of A^2/O-MBR, and the sustainable critical flux was 26 LMH.

Supplementary Materials: The following supporting information can be downloaded at: https://www.mdpi.com/article/10.3390/polym15102319/s1, Figure S1: R simulation regardless of CEB; Figure S2: R of simulation and measurement under different coefficients; Figure S3: R simulation and measurement of membrane module B and C. Table S1:Coefficients of different simulation process.

Author Contributions: Conceptualization, R.X. and Y.F.; methodology, R.X. and M.Y.; validation, R.X.; formal analysis, R.X.; data curation, R.X.; writing—original draft preparation, R.X.; writing—review and editing, R.X. and J.S.; visualization, R.X.; supervision, R.X.; project administration, R.X.; funding acquisition, Y.F. All authors have read and agreed to the published version of the manuscript.

Funding: This research was funded by the National Natural Science Foundation of China grant number [51278483] and the National Major Science & Technology Projects for Water Pollution Control and Management grant number [2012ZX07202-005].

Institutional Review Board Statement: Not applicable.

Data Availability Statement: Not applicable.

Conflicts of Interest: The authors declare no conflict of interest.

References

1. Le-Clech, P. Membrane bioreactors and their uses in wastewater treatments. *Appl. Microbiol. Biotechnol.* **2010**, *88*, 1253–1260. [CrossRef]
2. Shen, Y.X.; Xiao, K.; Liang, P.; Sun, J.Y.; Sai, S.J.; Huang, X. Characterization of soluble microbial products in 10 large-scale membrane bioreactors for municipal wastewater treatment in China. *J. Membr. Sci.* **2012**, *415*, 336–345. [CrossRef]
3. Gul, A.; Hruza, J.; Yalcinkaya, F. Fouling and Chemical Cleaning of Microfiltration Membranes: A Mini-Review. *Polymers* **2021**, *13*, 846. [CrossRef] [PubMed]
4. Chu, H.P.; Li, X.Y. Membrane fouling in a membrane bioreactor (MBR): Sludge cake formation and fouling characteristics. *Biotechnol. Bioeng.* **2005**, *90*, 323–331.
5. Wu, J.; Huang, X. Effect of mixed liquor properties on fouling propensity in membrane bioreactors. *J. Membr. Sci.* **2009**, *342*, 88–96. [CrossRef]
6. Meng, F.; Zhang, H.; Yang, F.; Zhang, S.; Li, Y.; Zhang, X. Identification of activated sludge properties affecting membrane fouling in submerged membrane bioreactors. *Sep. Purif. Technol.* **2006**, *51*, 95–103. [CrossRef]
7. Ji, L.; Zhou, J. Influence of aeration on microbial polymers and membrane fouling in submerged membrane bioreactors. *J. Membr. Sci.* **2006**, *276*, 168–177. [CrossRef]
8. Bouhabila, E.; Ben Aim, R.; Buisson, H. Fouling characterisation in membrane bioreactors. *Sep. Purif. Technol.* **2001**, *22*, 123–132. [CrossRef]
9. Sorm, R.; Bortone, G.; Saltarelli, R.; Jenicek, P.; Wanner, J.; Tilche, A. Phosphate uptake under anoxic conditions and fixed-film nitrification in nutrient removal activated sludge system. *Water Res.* **1996**, *30*, 1573–1584. [CrossRef]
10. Yao, M.; Ladewig, B.; Zhang, K. Identification of the change of soluble microbial products on membrane fouling in membrane bioreactor (MBR). *Desalination* **2011**, *278*, 126–131. [CrossRef]
11. Tian, Y.; Li, Z.; Ding, Y.; Lu, Y. Identification of the change in fouling potential of soluble microbial products (SMP) in membrane bioreactor coupled with worm reactor. *Water Res.* **2013**, *47*, 2015–2024. [CrossRef]
12. Bae, T.-H.; Tak, T.-M. Interpretation of fouling characteristics of ultrafiltration membranes during the filtration of membrane bioreactor mixed liquor. *J. Membr. Sci.* **2005**, *264*, 151–160. [CrossRef]
13. Meng, F.; Yang, F.; Shi, B.; Zhang, H. A comprehensive study on membrane fouling in submerged membrane bioreactors operated under different aeration intensities. *Sep. Purif. Technol.* **2008**, *59*, 91–100. [CrossRef]
14. Hung, T.S.; Bilad, M.R.; Shamsuddin, N.; Suhaimi, H.; Ismail, N.M.; Jaafar, J.; Ismail, A.F. Confounding Effect of Wetting, Compaction, and Fouling in an Ultra-Low-Pressure Membrane Filtration: A Review. *Polymers* **2022**, *14*, 2073. [CrossRef] [PubMed]
15. Field, R.W.; Wu, D.; Howell, J.A.; Gupta, B.B. Critical flux concept for microfiltration fouling. *J. Membr. Sci.* **1995**, *100*, 259–272. [CrossRef]
16. Le Clech, P.; Jefferson, B.; Chang, I.S.; Judd, S.J. Critical flux determination by the flux-step method in a submerged membrane bioreactor. *J. Membr. Sci.* **2003**, *227*, 81–93. [CrossRef]
17. Guglielmi, G.; Saroj, D.P.; Chiarani, D.; Andreottola, G. Sub-critical fouling in a membrane bioreactor for municipal wastewater treatment: Experimental investigation and mathematical modelling. *Water Res.* **2007**, *41*, 3903–3914. [CrossRef]
18. Li, J.; Zhang, X.; Cheng, F.; Liu, Y. New insights into membrane fouling in submerged MBR under sub-critical flux condition. *Bioresour. Technol.* **2013**, *137*, 404–408. [CrossRef]
19. Hwang, B.-K.; Lee, C.-H.; Chang, I.-S.; Drews, A.; Field, R. Membrane bioreactor: TMP rise and characterization of bio-cake structure using CLSM-image analysis. *J. Membr. Sci.* **2012**, *419*, 33–41. [CrossRef]

20. Nguyen, T.N.; Su, Y.C.; Pan, J.R.; Huang, C. Comparison of membrane foulants occurred under different sub-critical flux conditions in a membrane bioreactor (MBR). *Bioresour. Technol.* **2014**, *166*, 389–394. [CrossRef]
21. Xu, R.; Zhang, Q.; Tong, J.; Wei, Y.; Fan, Y. Internal carbon source from sludge pretreated by microwave-H_2O_2 for nutrient removal in A2/O-membrane bioreactors. *Environ. Technol.* **2014**, *36*, 827–836. [CrossRef] [PubMed]
22. Psoch, C.; Schiewer, S. Resistance analysis for enhanced wastewater membrane filtration. *J. Membr. Sci.* **2006**, *280*, 284–297. [CrossRef]
23. Schwarz, A.O.; Rittmann, B.E.; Crawford, G.V.; Klein, A.M.; Daigger, G.T. Critical review on the effects of mixed liquor suspended solids on membrane bioreactor operation. *Sep. Sci. Technol.* **2006**, *41*, 1489–1511. [CrossRef]
24. Wang, Z.; Wu, Z.; Yu, G.; Liu, J.; Zhou, Z. Relationship between sludge characteristics and membrane flux determination in submerged membrane bioreactors. *J. Membr. Sci.* **2006**, *284*, 87–94. [CrossRef]
25. Janus, T.; Paul, P.; Ulanicki, B. Modelling and simulation of short and long term membrane filtration experiments. *Desalination Water Treat.* **2009**, *8*, 37–47. [CrossRef]
26. Okamura, D.; Mori, Y.; Hashimoto, T.; Hori, K. Identification of biofoulant of membrane bioreactors in soluble microbial products. *Water Res.* **2009**, *43*, 4356–4362. [CrossRef]
27. Li, X.Y.; Wang, X.M. Modelling of membrane fouling in a submerged membrane bioreactor. *J. Membr. Sci.* **2006**, *278*, 151–161. [CrossRef]
28. Metcalf, L. *Wastewater Engineering: Treatment and Reuse*; McGraw-Hill: New York, NY, USA, 2003.
29. Li, X.Y.; Chu, H.P. Membrane bioreactor for the drinking water treatment of polluted surface water supplies. *Water Res.* **2003**, *37*, 4781–4791. [CrossRef]
30. Sørensen, B.L.; Sørensen, P.B. Structure compression in cake filtration. *J. Environ. Eng.* **1997**, *123*, 345–353. [CrossRef]
31. Boyle-Gotla, A.; Jensen, P.D.; Yap, S.D.; Pidou, M.; Wang, Y.; Batstone, D.J. Dynamic multidimensional modelling of submerged membrane bioreactor fouling. *J. Membr. Sci.* **2014**, *467*, 153–161. [CrossRef]
32. Li, X.Y.; Yuan, Y. Collision Frequencies of Microbial Aggregates with Small Particles by Differential Sedimentation. *Environ. Sci. Technol.* **2002**, *36*, 387–393. [CrossRef] [PubMed]
33. Feng, S.; Zhang, N.; Liu, H.; Du, X.; Liu, Y.; Lin, H. The effect of COD/N ratio on process performance and membrane fouling in a submerged bioreactor. *Desalination* **2012**, *285*, 232–238. [CrossRef]
34. Qing, Z.; Rongle, X.; Xiang, Z.; Yaobo, F. Simulation and optimization of airlift external circulation membrane bioreactor using computational fluid dynamics. *Water Sci. Technol.* **2014**, *79*, 1846–1852. [CrossRef] [PubMed]
35. Krzeminski, P.; Iglesias-Obelleiro, A.; Madebo, G.; Garrido, J.M.; van der Graaf, J.H.J.M.; van Lier, J.B. Impact of temperature on raw wastewater composition and activated sludge filterability in full-scale MBR systems for municipal sewage treatment. *J. Membr. Sci.* **2012**, *423*, 348–361. [CrossRef]
36. Jiang, T.; Kennedy, M.D.; Guinzbourg, B.F.; Vanrolleghem, P.A.; Schippers, J.C. Optimising the operation of a MBR pilot plant by quantitative analysis of the membrane fouling mechanism. *Water Sci. Technol.* **2005**, *51*, 19–25. [CrossRef] [PubMed]
37. Van den Brink, P.; Satpradit, O.A.; van Bentem, A.; Zwijnenburg, A.; Temmink, H.; van Loosdrecht, M. Effect of temperature shocks on membrane fouling in membrane bioreactors. *Water Res.* **2011**, *45*, 4491–4500. [CrossRef]
38. Wedi, D.; Bleisteiner, S.; Wild, W. Seasonal changes in filterability and permeability shown by the example of MBR-Monheim. In Proceedings of the 8 Aachener Tagung Wasser und Membrane, Aachen, Germany, 27–28 October 2009; pp. A26-1–A26-11.
39. Martín-Pascual, J.C.L.-D.J.; López-López, C.; Muñio, M.M.; Hontoria, E.; Poyatos, J.M. Effects of temperature on the permeability and critical flux of the membrane in a moving bed membrane bioreactor. *Desalination Water Treat.* **2013**, *53*, 3439–3448. [CrossRef]
40. Fenu, A.; De Wilde, W.; Gaertner, M.; Weemaes, M.; de Gueldre, G.; Van De Steene, B. Elaborating the membrane life concept in a full scale hollow-fibers MBR. *J. Membr. Sci.* **2012**, *421*, 349–354.
41. Guglielmi, G.; Chiarani, D.; Judd, S.; Andreottola, G. Flux criticality and sustainability in a hollow fibre submerged membrane bioreactor for municipal wastewater treatment. *J. Membr. Sci.* **2007**, *289*, 241–248. [CrossRef]
42. Zsirai, T.; Buzatu, P.; Aerts, P.; Judd, S. Efficacy of relaxation, backflushing, chemical cleaning and clogging removal for an immersed hollow fibre membrane bioreactor. *Water Res.* **2012**, *46*, 4499–4507. [CrossRef]

polymers

MDPI

Article

Influence of Curing Agent Amount on Properties of Dynamic Vulcanized Phenyl Silicone Rubber-SEBS-SBS System

Chunxu Zhao, Bobing He * and Xian Chen *

College of Chemistry, Sichuan University, Chengdu 610065, China
* Correspondence: hebobing@scu.edu.cn (B.H.); chen.xian@scu.edu.cn (X.C.); Tel.: +86-028-85468166 (B.H. & X.C.)

Abstract: In this paper, we prepared a new type of thermoplastic vulcanizate (TPV) by melt blending methyl vinyl phenyl silicone rubber (PSR), styrene butylene copolymer (SBS), and hydrogenated SBS (SEBS) and then dynamically vulcanizing it. At the same time, we studied the influence of the content of the vulcanizing agent on the properties. The corresponding backscattered electron images were obtained by a scanning electron microscope (SEM) test of each group of samples, as well as the distribution of the PSR phase and the SEBS-SBS phase, and the vulcanization process of the samples with a vulcanizing agent content of 1 phr were characterized. According to the imaging principle of the backscattered electron signal, we found that the atomic number contrast can be clearly reflected in the backscattered image. From the obtained images, we found that PSR is a dispersed phase, while SEBS and SBS are continuous phases, that is, they had a "Sea-Island" structure. In the first 30 s of the vulcanization reaction, the "Sea-Island" structure is formed, and then the vulcanization reaction rate gradually slows down. We then printed the images and analyzed them using a colorimeter and found that it was feasible to quantitatively characterize the size of the compatible layer between the continuous and dispersed phases. According to the quantitative characterization results, we found that the silane coupling agent KH-172 can increase the thickness of the compatible layer by nearly 35%. In addition, we also tested the mechanical properties and low-temperature elastic properties of the material. Finally, we found that when the content of the vulcanizing agent was 1 phr, the elastic properties and tensile properties were the best, and when the content of the vulcanizing agent was more than 1 phr, the tensile and elastic properties of the material decreased significantly. At the same time, we also found that the addition of the silane coupling agent KH-172 can also significantly improve the tensile properties and elastic properties of TPV, which we believe is related to the increase in the thickness of the compatible layer. The test results of dynamic mechanics show that PSR has good compatibility with SEBS-SBS. When the vulcanizing agent content is less than or equal to 1 phr, the material exhibits good low-temperature resistance. In addition, through the test of the melt index of each group, it was also found that the addition of the vulcanizing agent will affect the fluidity of the melt to a certain extent. When the content of the vulcanizing agent is greater than 1 phr, the melt fluidity decreases more obviously.

Keywords: thermoplastic vulcanized rubber; methyl vinyl phenyl silicone rubber; backscattered electrons; dynamic vulcanization; scanning electron microscope

Citation: Zhao, C.; He, B.; Chen, X. Influence of Curing Agent Amount on Properties of Dynamic Vulcanized Phenyl Silicone Rubber-SEBS-SBS System. *Polymers* **2022**, *14*, 5443. https://doi.org/10.3390/polym14245443

Academic Editor: Andrea Sorrentino

Received: 18 September 2022
Accepted: 9 December 2022
Published: 12 December 2022

1. Introduction

Silicone rubber is a polyorganosiloxane whose main chain is composed of silicon and oxygen. However, silicone rubber is divided into different types, according to the branched chain [1]. Figure 1 shows the structural formula of methyl vinyl phenyl silicone rubber (PSR). According to the ratio of the number of moles of phenyl-containing chain units to the number of moles of all chain units, we often designate phenyl silicone rubber with a mole fraction less than 10% as low phenyl silicone rubber [2]. It has attracted much attention because of its good elasticity and low-temperature resistance [3,4]. However, its application conditions are limited due to its poor mechanical strength [5]. Blending silicone rubber with other polymers is a common modification method [6–8]; thus, we plan to prepare a modified thermoplastic elastomer (TPE) by blending silicone rubber with SBS and SEBS through dynamic vulcanization. Generally, TPE prepared by dynamic vulcanization is called thermoplastic vulcanized rubber (TPV) [9]. The first developed TPV is made of ethylene propylene diene monomer (EPDM) and polypropylene (PP), and it was the first to start the industrial production of this TPV by Monsanto [10–12]. SBS is a block copolymer made from the polymerization of styrene and butadiene. It is often used as a modifier of asphalt and has good tensile strength, low-temperature performance, and processability [13,14]. SEBS is the product of SBS hydrogenation; hence, in addition to the advantages of SBS, SEBS also has good aging resistance, ozone resistance, and other properties [15–17]. In this paper, the modified TPV we prepared is a new material with the advantages of these three polymers. In the preparation process of this material, process conditions, such as temperature and rotor speed, and the content of the different substances in the formulation will affect the performance of the TPV [18,19]. However, since we prepare materials by dynamic vulcanization, and vulcanization is an indispensable step to impart performance to rubber [20], it is very important to study the effect of the vulcanizing agent content on the TPV performance and find the appropriate vulcanizing agent content.

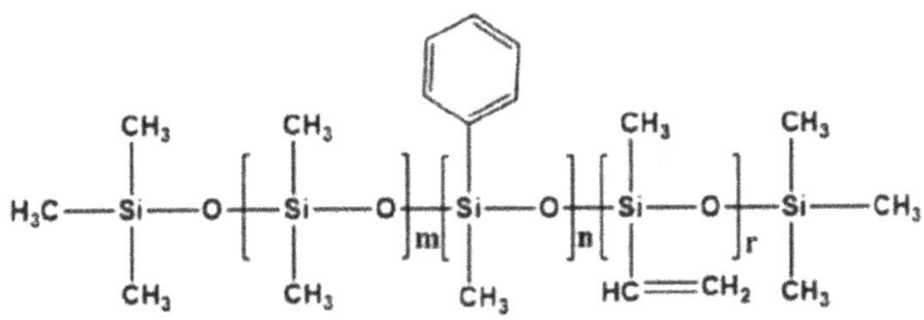

Figure 1. The structural formula of methyl vinyl phenyl silicone rubber.

As mentioned above, our method for preparing this novel modified TPV is dynamic vulcanization. This is a process in which rubber and thermoplastics are melted and blended in a high-temperature, high-shear mixer, and the vulcanized rubber is broken up by high-shear force to disperse it well in the thermoplastic component [21,22]. Therefore, it is necessary to observe the distribution among the phases and find the correlation with the macroscopic properties. Usually, for the phase distribution study of polymer blends, researchers often use the etching method. This is a method of operation that dissolves a phase in a polymer blend and takes a secondary electron signal map of the undissolved phase [23]. In this way, we see to some extent the distribution of the polymer and the interaction between the phases. However, this method has the disadvantages that it is difficult to find a good solvent for a certain phase and it is difficult to determine the complete dissolution [24]. Therefore, to better characterize the distribution of different phases, in this paper, we use the image generated by the backscattered electron signal of the scanning electron microscope to characterize the phase distribution, and we also creatively characterize the thickness of the compatibility layer between the two phases quantitatively. Since the backscattered electron signal is related to the atomic number [25–27], the characterization method of the backscattered electron signal map is often used to observe the phase change of the alloy [28,29] and the phase distribution of the cement section [30,31]. When this characterization method is applied to the observation of polymer blends, for polymer blend systems with no difference in atomic number, researchers often dye the samples to make the characterization results of the backscattered electron signal clear [32,33]. There are also reports on the characterization of the three-dimensional structure of polymer blends using this method [34]. In this case, however, there are differences in atomic numbers between the components in the blend, thus we can obtain good results without dyeing. In addition, backscattered electron images can also overcome the shortcomings of secondary electron image characterization methods, such as difficulty finding a good solvent for the etched phase and the need to destroy the sample for testing [35,36].

2. Experimental

2.1. Materials

Methyl vinyl phenyl silicone rubber (LE-1150, Ron Silicon Material Co., Ltd., Nanjing, China), SEBS (YH-602T, Baling Petrochemical Branch of Sinopec Group, Yueyang, China), SBS (D1155JOP, Kraton, Belpre, OH, USA), white oil (250N, Wanghai Petrochemical Co., Ltd., Taizhou, China), silane coupling agent KH-172 (vinyltris(2-methoxyethoxy)silane, Yuanjin New Material Co., Ltd., Qufu, China), platinum catalyst (Daxi Chemical Raw Materials Co., Ltd., Guangzhou, China), and hydrogen-containing silicone oil (hydrogen content 1.6%, Xinglongda New Materials Co., Ltd., Jinan, China).

2.2. Sample Preparation

According to the formulations shown in Table 1, the samples differed only in the hydrogen-containing silicone oil content. Before adding the sample to the torque rheometer, we need to use an overhead electric mixer (Kexing Instrument Co., Ltd., Shanghai, China) to pre-mix the SBS, SEBS, and white oil (the total volume of the mixture is about 190 mL) at a speed of 100 rpm, and then let stand for 30 min, in order for the white oil to be more fully absorbed by the SBS and SEBS. Subsequently, we added the previous mixture into the torque rheometer and added phenyl silicone rubber after the torque value was stable. The temperature in the torque rheometer chamber was 170 °C, and the rotor speed was 70 rpm. After the torque value reaches equilibrium again, we sequentially add a silane coupling agent, hydrogen-containing silicone oil, and platinum catalyst to let the vulcanization reaction start until the torque value reaches equilibrium, which means that the vulcanization reaction is basically over. After the vulcanization reaction, we take out the mixture from the mixing chamber. Then we put the mixture into a thermocompression molding machine for processing, and after processing, a sheet with a size of 146 × 144 × 2 mm^3 was obtained for further experiments.

Table 1. TPV dynamic vulcanization experimental formulations (unit: phr).

Entry	PSR	SEBS	SBS	250N	KH-172	Platinum Catalyst	Hydrogen-Containing Silicone Oil
1	50	25	25	10	1	0.5	0
2	50	25	25	10	1	0.5	0.5
3	50	25	25	10	1	0.5	1
4	50	25	25	10	1	0.5	1.5
5	50	25	25	10	1	0.5	2

2.3. Scanning Electron Microscope Test

A total of 5 rectangular specimens (dimensions of 10 × 2 × 2 mm^3) were cut out from the 5 sheets prepared before, and then we clamped them with tweezers and placed them in liquid nitrogen to freeze for 3 min. We then took it out, broke it immediately, and sprayed gold on the cross-section after the sample was broken. The gold-sprayed samples could be imaged using a scanning electron microscope (Thermo Fisher Scientific, Helios G4 UC, Waltham, MA, USA). When observing the vulcanization process, it is not necessary to press the molten blend into tablets. To ensure the accurate vulcanization time of the characterization, we immediately took out the blend from the torque rheometer cavity and put it into liquid nitrogen for cooling. After quenching and spraying gold, the corresponding images were observed.

2.4. Fourier Transform Infrared Spectroscopy Test

We took a sample of size 10 × 8 × 2 mm^3 from the prepared sheet and processed the sample using the tableting method. After the sample was processed, it was tested by Fourier transform infrared spectroscopy (Bruker Scientific Technology Co., Ltd., Beijing, China). The relevant parameters are: the number of scans is 16, and the detection mode is ATR with a resolution of 4 cm^{-1}.

2.5. Compatibility Layer Quantitative Characterization Test

The 2000-times backscattered electron image (image pixel: 3840 × 2160 dpi) obtained by the scanning electron microscope test was printed on photo paper by a high-definition printer. The pixel of the printer is 1200 × 1200 dpi, and the size of the photo paper is 420 mm × 297 mm. We first tested the color space coordinates of the bright and dark parts of the image. Then we put the light port of the colorimeter in the dark place of the backscattered image; moved the colorimeter in the direction shown in Figure 2, moving 1 mm each time; and recorded the color space coordinate values here for further analysis of the data. Generally, the faster the color space coordinate value changes, the smaller the thickness of the compatible layer and the weaker the interaction between the two phases. It should be noted that to reduce the test error, at least four directions should be tested for each disperse phase tested.

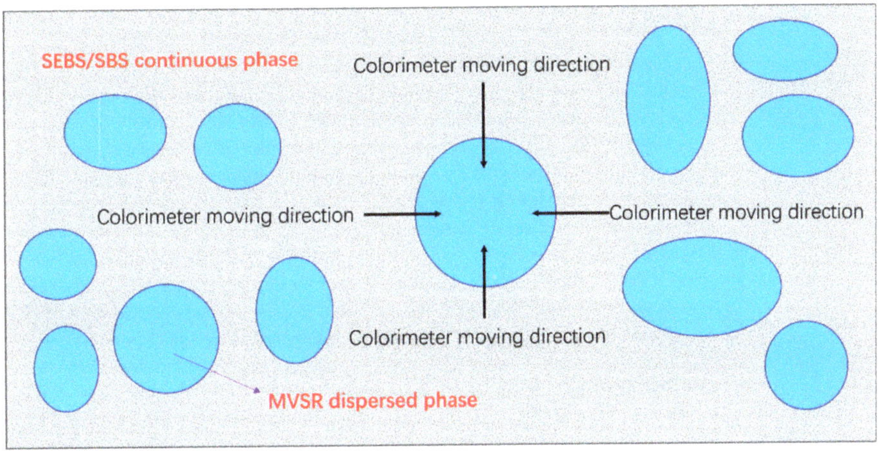

Figure 2. The test method for the colorimeter to test the size of the compatibility layer in TPV.

2.6. Mechanical Property Test

We used a dumbbell-shaped cutter to cut the original sample. After cutting, we obtained 5 dumbbell-shaped samples with a gauge length of 20 mm, a width of 4 mm, and a thickness of 2 mm. The reference standard is GB/T 528-2009. The mechanical properties of each dumbbell-shaped specimen were then tested using a tensile testing machine (AGS-J, Shimadzu Corporation, Kyoto, Japan), where the tensile rate was 200 mm min^{-1}. To test the tensile set of the experiment, we placed the broken specimens for about 5 min, after which the gauge length of the specimens was measured. The arithmetic means of the values of each group of mechanical property tests were taken as the result.

2.7. Dynamic Mechanical Properties Test

We cut out rectangular samples with a length of 35 mm, a width of 8 mm, and a thickness of 2 mm from the 5 sets of samples prepared in Section 2.2. The storage modulus and tangent value of the loss angle of the sample at different temperatures were tested by a dynamic thermomechanical analyzer (Q850, Walters Technology Co., Ltd., Shanghai,

China), in which the test temperature range was −100 °C to 0 °C, and the heating rate was 5 °C min^{-1}. The type of clamp used is the tensile clamp.

2.8. Low-Temperature Hardness Test

We put the 5 groups of samples prepared in Section 2.2 into an ultra-low temperature refrigerator, adjusted the temperature of the refrigerator to −60 °C, and then froze for 72 h. After that, the hardness test was carried out, and the hardness tester was placed in the refrigerator for 30 min before the test. Each group of samples tested 10 groups of data to take the arithmetic mean. In order to prevent a temperature rise in the refrigerator caused by continuous testing, which could affect the test results, we put the hardness tester into the refrigerator after testing 3 sets of data, closed the refrigerator door, and let it stand for 10 min before subsequent testing.

2.9. Melt Index Test

We took the melt blend prepared according to Table 1 out of the torque rheometer and put it into a melt flow rate tester (XRN-400C, Jinjian Testing Instrument Co., Ltd., Chengde, China) to preheat for 10 min. After, we started the melt index test. The test temperature was 190 °C and the load weight was 5 kg.

3. Results and Discussions

3.1. Torque Variation Diagram and Infrared Spectrum Analysis of Modified TPV Made of PSR and SEBS-SBS with Different Vulcanizing Agent Contents

The torque variation diagram of samples 1–5 is shown in Figure 3. Sample 1 is a simple blended TPV control group without a vulcanizing agent. When no vulcanizing agent is added, the torque value is balanced around 5.5 Nm. After adding the vulcanizing agent, it is not difficult to see that with increasing amounts of vulcanizing agent added, both the torque value and the equilibrium torque value increase after vulcanization. Because the vulcanization reaction is a cross-linking reaction that occurs inside the polymer, which is microscopically manifested as the formation of a three-dimensional network structure, the torque value increasing speed can be regarded as a characterization of the intensity of the cross-linking reaction inside the torque rheometer. From this, we see that the experimental group with the addition of 2 phr vulcanizing agents has the most severe cross-linking reaction, but as the increase of vulcanizing agent is equal between each group, we also find that the increased rate of the torque graph rise rate is not uniform. When the content of the vulcanizing agent increased from 0 phr to 1 phr, the increase of the cross-linking reaction rate was the most obvious, while the increase of the cross-linking reaction rate decreased when the content of the vulcanizing agent increased from 1 phr to 2 phr. In addition, we also find that with the increase of the vulcanizing agent content, the time for the vulcanization system to reach torque equilibrium will decrease, that is, the time of the vulcanization reaction decreases.

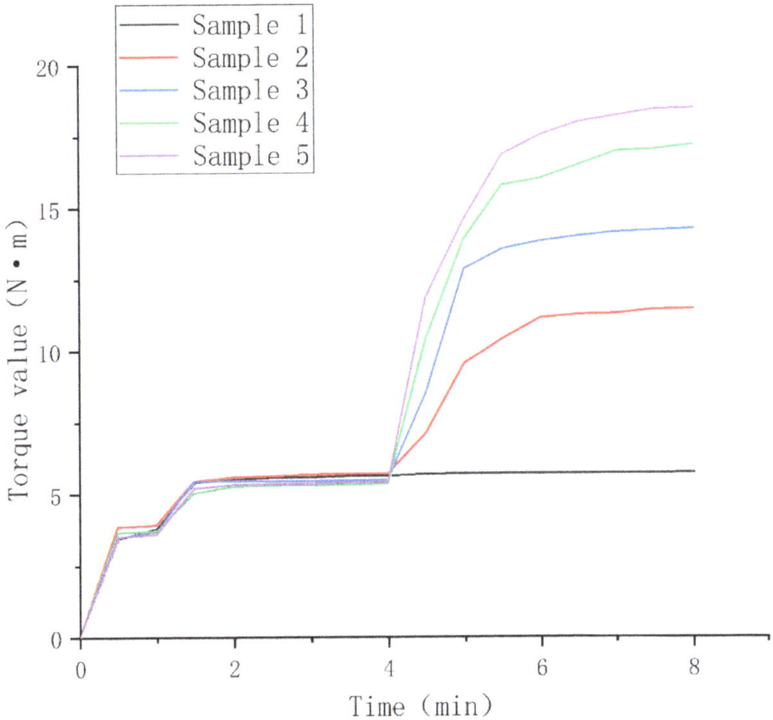

Figure 3. Torque value change diagram of samples 1–5 during vulcanization.

In addition, we carried out infrared spectroscopy tests on samples 1–5, and the test results are shown in Figure 4. From Table 2, we know that the wavenumbers of 910 cm^{-1} and 966 cm^{-1} correspond to the characteristic absorption peaks of the 1,2 chain unit and the characteristic absorption peak of the anti-1,4 chain unit, respectively. As the essence of hydrosilylation was a hydrogenation reaction of double bonds, we found that the transmittance of each group of samples at wave numbers 910 cm^{-1} and 966 cm^{-1} as in Figure 4B was significantly different. When the content of the vulcanizing agent is 0.5–1.5 phr, the degree of hydrogenation of the 1,2 chain segment has almost no change, and only after further addition of the vulcanizing agent can the degree of hydrogenation at the 1,2 chain segment be increased. The situation at 966 cm^{-1} is completely different from that at 910 cm^{-1}. When the vulcanizing agent content increases from 0.5 phr to 1.5 phr, the degree of hydrogenation of the anti-1,4 chain units increases to varying degrees. Interestingly, after further increasing the amount of the vulcanizing agent, the degree of hydrogenation does not change, but hydrogenation occurs at the 1,2 chain link. In general, increasing the amount of vulcanizing agent can make the hydrogenation degree of this TPV system continue to increase, but there is a difference in the hydrogenation position.

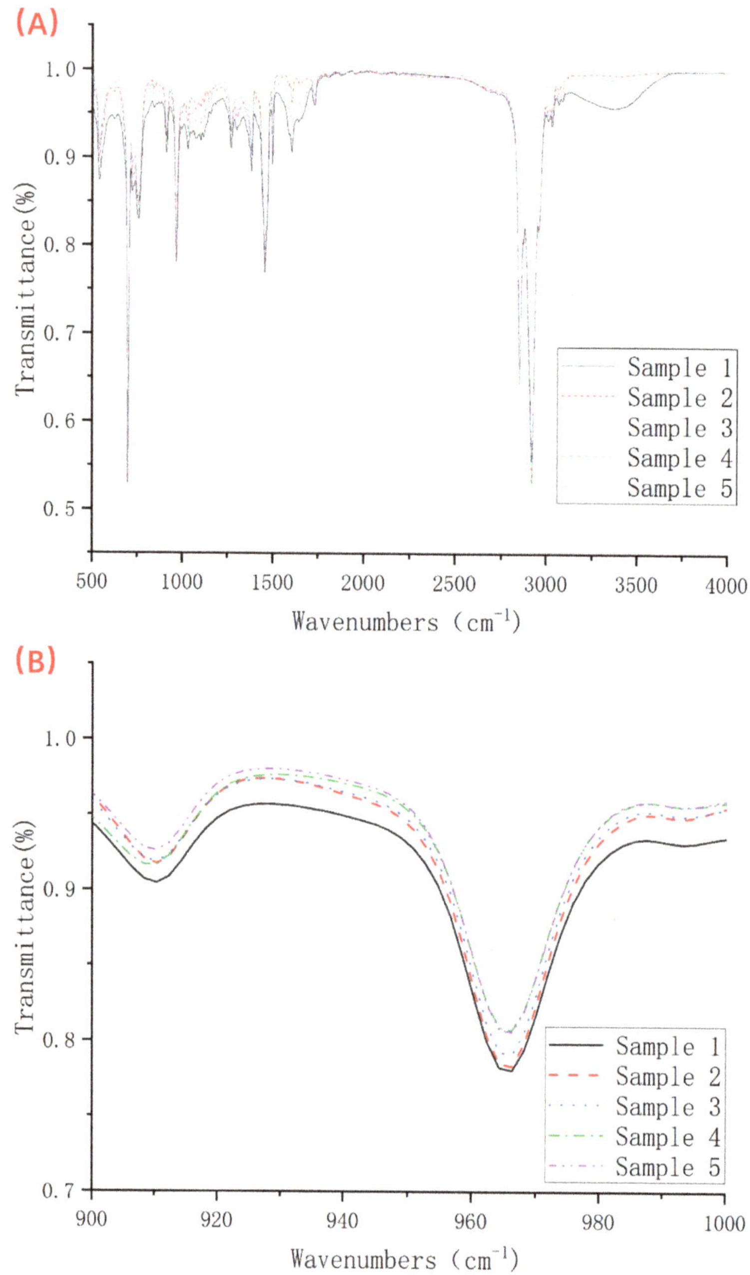

Figure 4. Infrared spectra of samples 1–5 (where (**A**) is the complete spectrum and (**B**) is the detailed image of the wave number 900–1000 cm^{-1}).

Table 2. Characteristic peaks, corresponding groups, and vibration types of SBS.

Wavenumber (cm^{-1})	Designation of IR Spectra
541	Styrene block, deformations of the aromatic ring
700, 756	Out of plane deformations of the aromatic ring
720	-(CH_2)- rocking vibration, -(CH_2)$_n$-, n, number of consecutive -CH_2- groups are more than 4 in SEBS, the same as in polyethylene
738	Out of plane deformation of cis double bonds overlapped by peak 756 cm^{-1} from an aromatic ring
910	δ-$_{C-H}$, =C-H, out of plane deformations of vinyl units
966	Out of plane deformation of trans-unit
1379	-$CH(CH_2CH_3)CH_2$- unit, -CH_2 for hydrogenated products SEBS
1453	Deformation of aromatic ring
1470	Asymmetric vibrations and deformations of groups -CH_2, -CH_3 in SEBS
1492, 1602	Stretching vibration of the double bond of the aromatic ring
1641	Stretching vibration of the double bond of vinyl units

3.2. The Backscattered Electron Images of TPV Made of PSR and SEBS-SBS with Different Vulcanizing Agent Contents and Analysis of the Characterization Results of the Vulcanization Process

Scanning electron microscopy is a characterization method that obtains information about the surface of a sample by analyzing the electronic signal excited by the interaction of incident electrons with the sample. Etching, which is commonly used for the microscopic characterization of polymeric materials, uses the secondary electron signal, while the scanning electron microscope images taken in this paper use the backscattered electron signal. As the intensity of the backscattered electron signal is related to the atomic number, when the atomic number is less than 20, the relationship between the signal intensity and the atomic number is almost linear. Therefore, the silicon-dominated PSR phase and the carbon-dominated SEBS-SBS phase in TPV can be clearly distinguished in the backscattered images [37].

Figure 5 shows the 2000-times backscattered electron images of samples 1–5. According to the above-mentioned backscattered electron imaging principle, we clearly find that the PSR phase and the SEBS-SBS phase have a clear degree of distinction. From Figure 5A, we found that the PSR phase and the SEBS-SBS phase in the simple blend TPV without vulcanizing agent were in a co-continuous state, and it was impossible to clearly distinguish which was a continuous phase and which was a dispersed phase, which was particularly obvious in the right half of the image. When the amount of vulcanizing agent added was increased to 0.5 phr, a "Sea-Island" structure began to appear between the PSR phase and the SEBS-SBS phase, in which the PSR phase was dispersed as a dispersed phase between the continuous phases composed of the SEBS-SBS phase. However, interestingly, from the middle and bottom right of Figure 5B, we see that the dispersion of the PSR phase in the SEBS-SBS phase is not very uniform. When the amount of vulcanizing agent is increased to 1 phr, we find from Figure 5C that the dispersion uniformity of the PSR phase in the SEBS-SBS phase is significantly better than that of sample 2, and the "Sea-Island" structure is also clearer. However, when we continued to increase the content of the vulcanizing agent, we were surprised to find that the dispersion of the dispersed phase became significantly worse. Through Figure 5D,E, we see that a large number of PSR phases agglomerate together, which makes the co-continuous structure between the PSR phase and the SEBS-SBS phase more inclined. This shows that the vulcanizing agent has a certain promoting effect on the formation of the "Sea-Island" structure in TPV, but this effect does not increase with the increase of the vulcanizing agent content. In this experiment, the vulcanizing agent content of 1 phr is the most suitable.

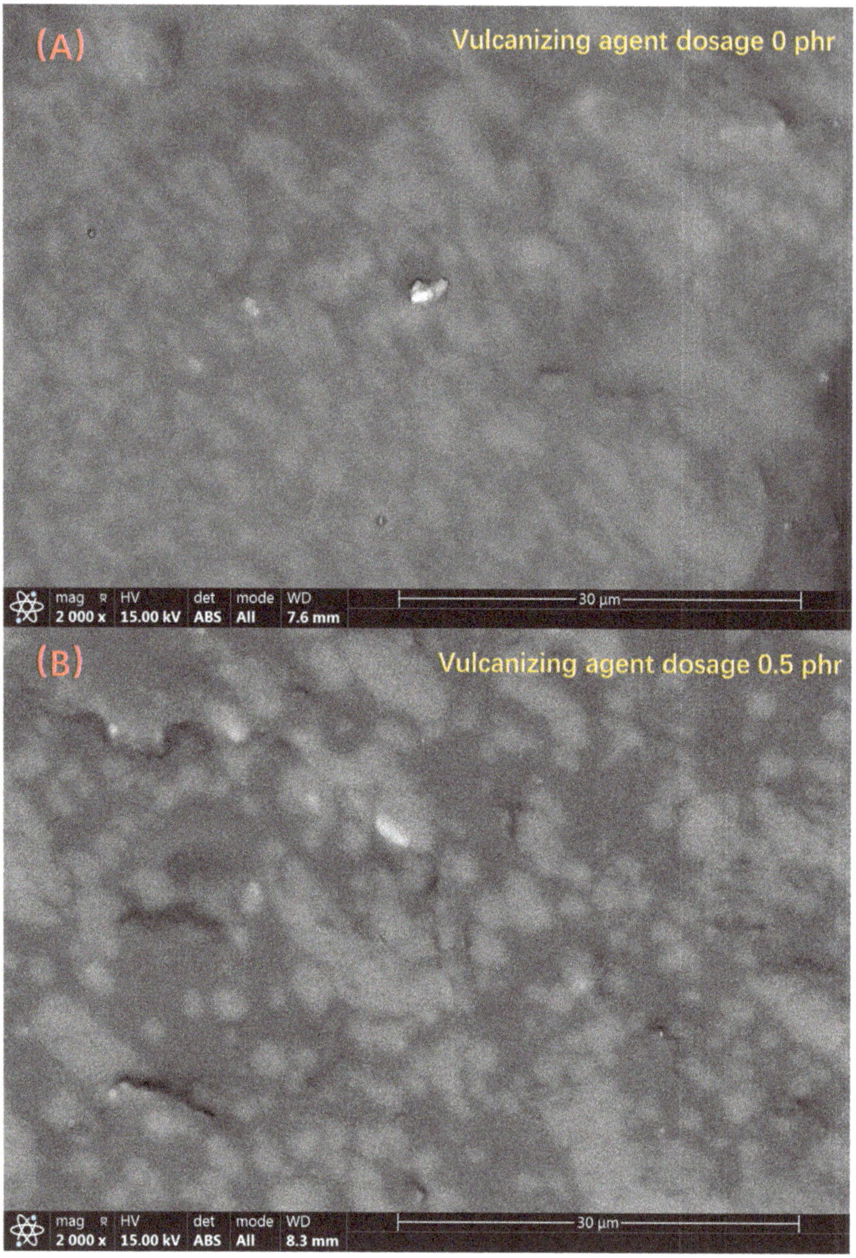

Figure 5. *Cont.*

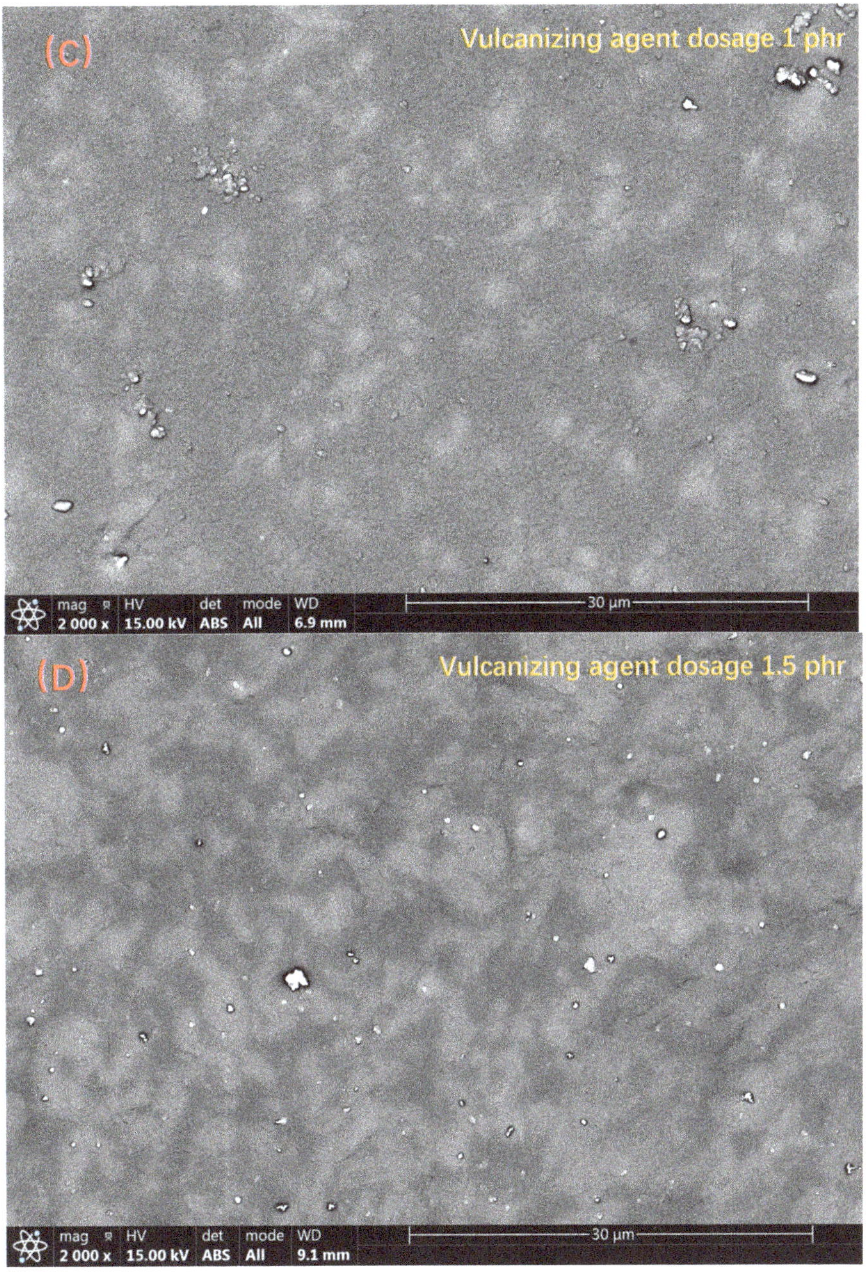

Figure 5. *Cont.*

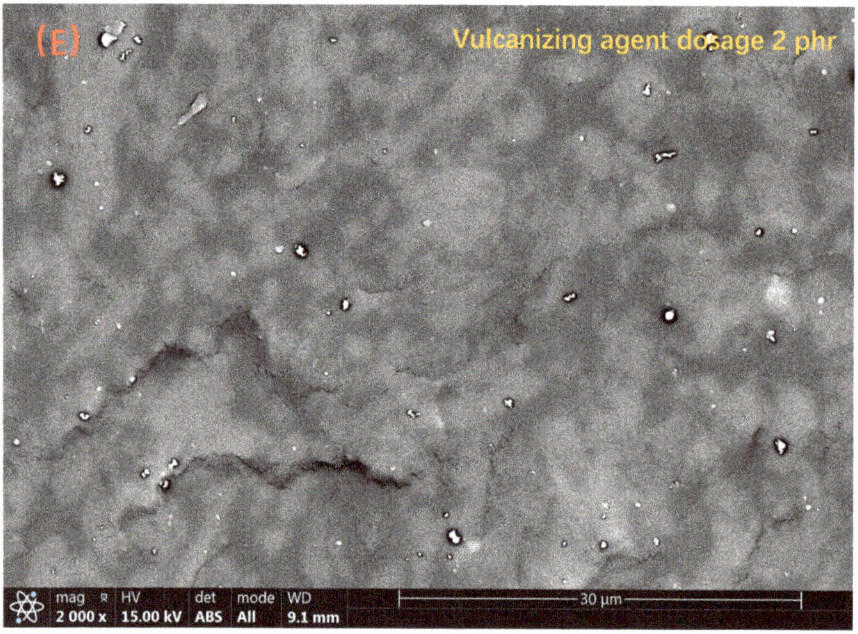

Figure 5. Backscattered electron images of samples 1–5 ((**A–E**) correspond to samples 1–5, respectively).

Next, we characterize the formation process of the "Sea-Island" structure. As shown in Figure 6, the 1000-times backscattered images of sample 3 taken at different curing times, we see the formation process of the "Sea-Island" structure. Initially, when no vulcanizing agent was added, as shown in Figure 6A, the PSR phase and the SEBS-SBS phase exhibited a co-continuous structure, and we could not distinguish the continuous phase from the dispersed phase. When the vulcanization reaction occurs for 0.5 min, according to the change of torque value in Figure 3, we see that the torque increases rapidly, which means that the initial reaction of the vulcanization reaction is rapid. Combined with Figure 6B, we found that the change at the microscopic level corresponding to the torque value at this time is the formation of a "Sea-Island" structure, in which the PSR phase is the dispersed phase and the SEBS-SBS phase is the continuous phase. After, with the progress of the vulcanization process, when the vulcanization reaction occurred for 1–1.5 min, the smaller diameter PSR phase was more evenly distributed in the SEBS-SBS phase, but the larger diameter PSR phase marked in Figure 6C,D still existed. When the vulcanization reaction continues, we see that the diameter of the PSR phase decreases further through the marks in Figure 6D,E, and the PSR phase with a larger diameter almost disappears in the observation field. According to Figure 3, the torque value change tends to be stable at this time, which means that the vulcanization reaction tends to end. The reason for the smaller diameter of the PSR phase and the more uniform dispersion is that the rotation of the rotor applies an external shear force to the blend such that the dispersed phase is more uniformly dispersed in the continuous phase.

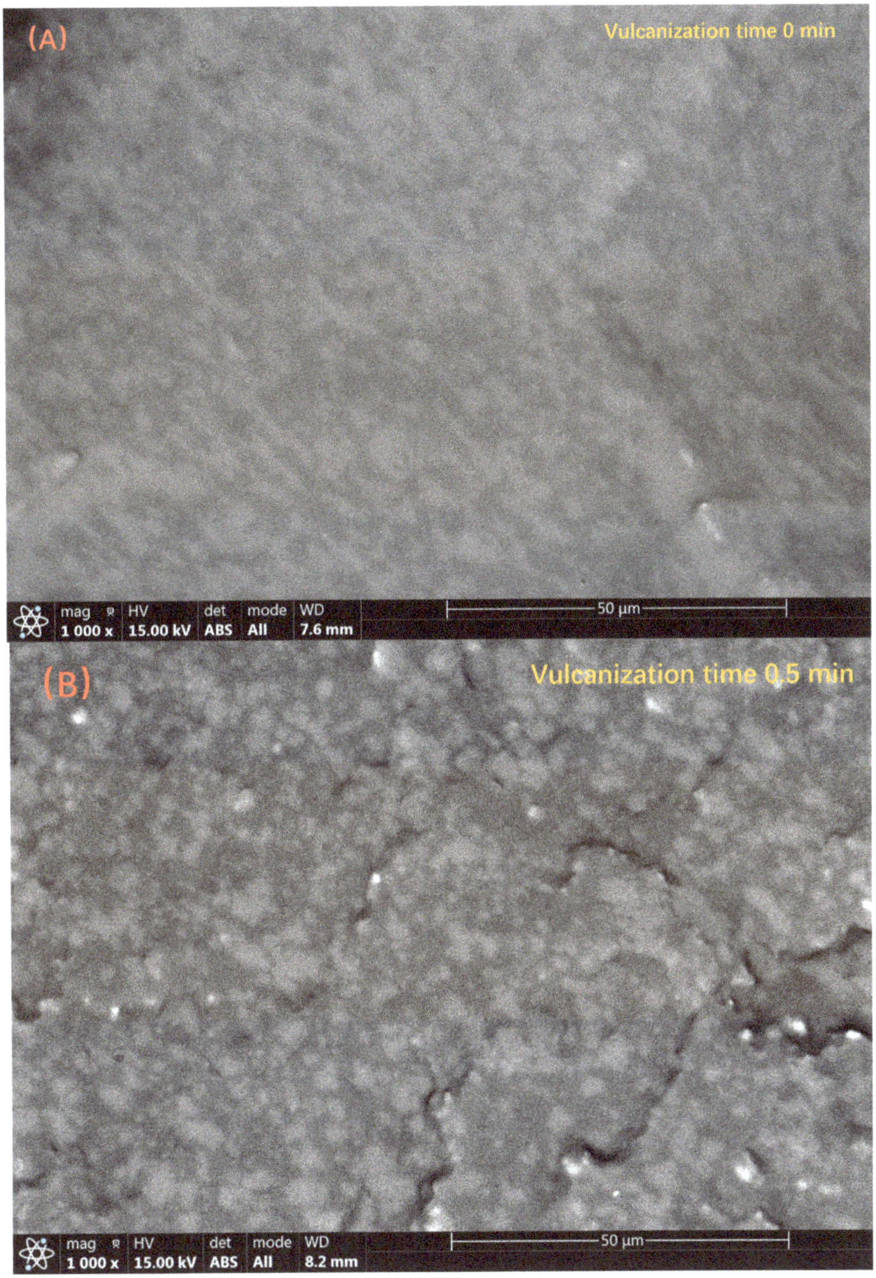

Figure 6. *Cont.*

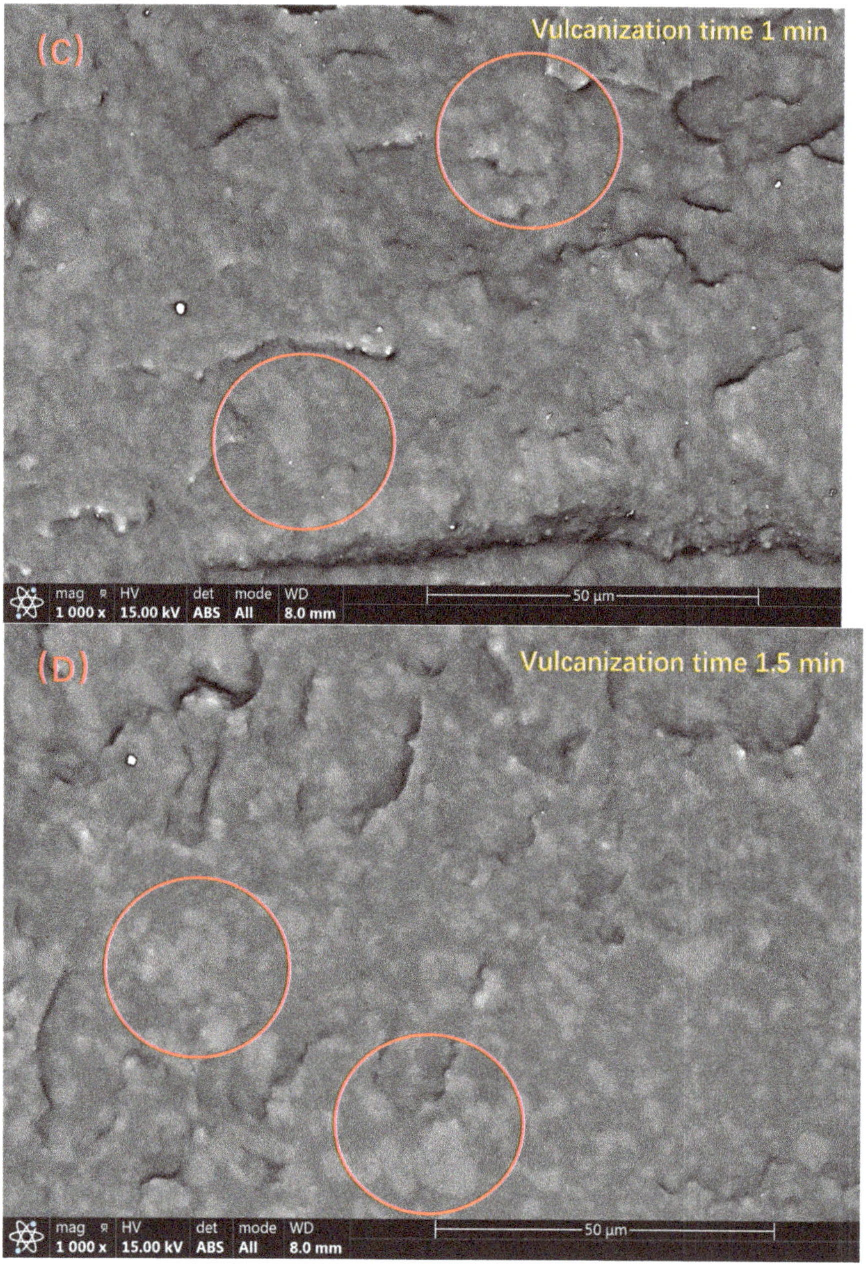

Figure 6. *Cont.*

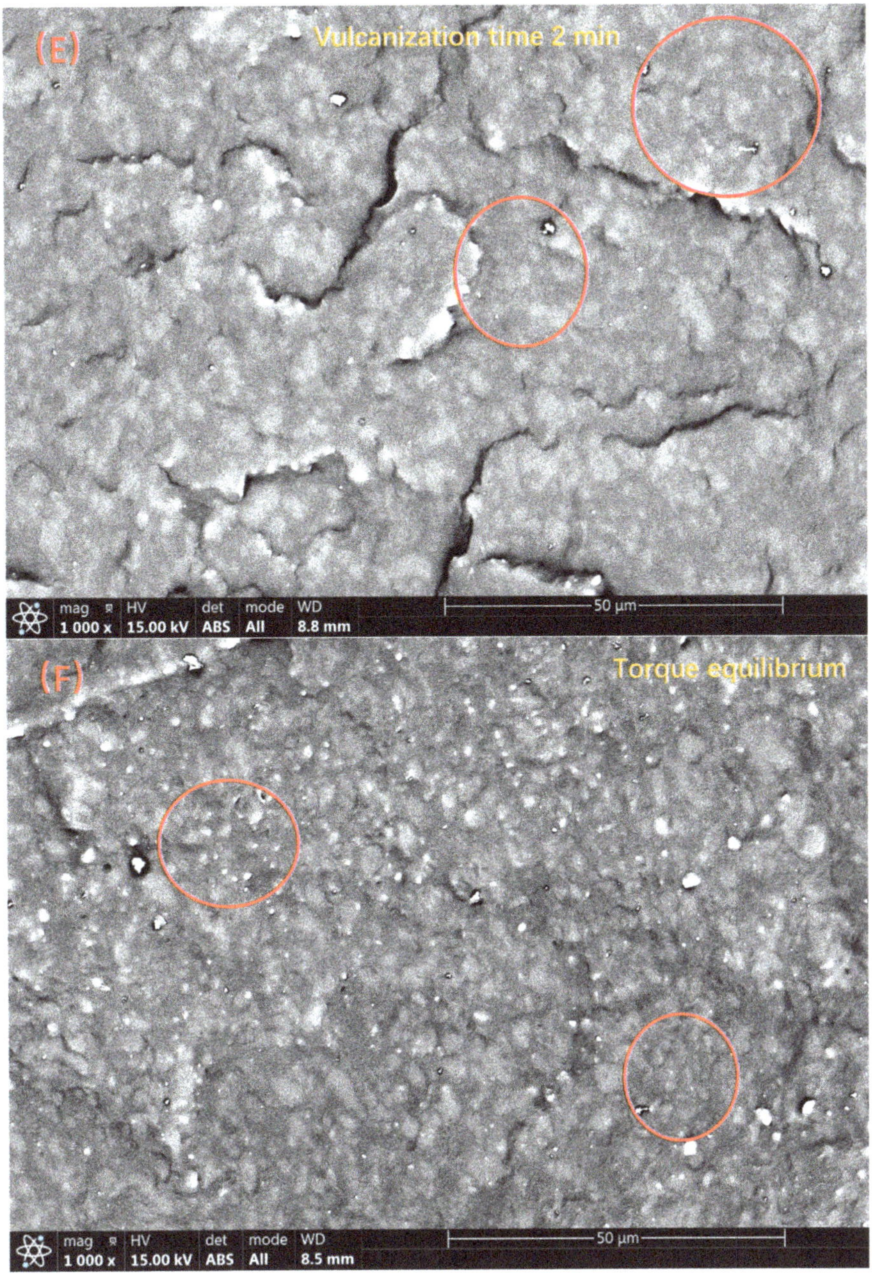

Figure 6. The backscattering image of sample 3 during the vulcanization process ((**A**–**E**) correspond to the vulcanization times of 0 min, 0.5 min, 1 min, 1.5 min, and 2 min, respectively, and (**F**) is the backscattered image after torque equilibrium. The change in the particle size of PSR as the vulcanization process proceeds is marked by the circles in the figure).

3.3. Analysis of the Results of the Size Chart of the Compatibility Layer and Mechanical Property Test Results

We often use color space coordinates to describe different colors, and each color has a one-to-one correspondence with its color space coordinates. This coordinate consists of the three variables *L*, *A*, and *B*, representing black and white, red-green, and yellow-blue, respectively. In this paper, the backscattered electron image we obtained is black and white; hence, we only use the *L* value to describe the color of different positions in the image. To quantitatively characterize the thickness of the compatible layer, we tested the *L* value of the continuous phase and the dispersed phase and found that the *L* value of the continuous phase was 70–72, and the *L* value of the dispersed phase was 47–50. Therefore, we take the distance with an *L* value of 50–70 as the dimension of the compatibility layer.

We printed 2000-times BSE images on A3 size (420 mm × 297 mm) photo paper with a high-definition printer and moved the colorimeter as shown in Figure 1. Through the scale conversion of backscattered images, we calculated that the field of view of each backscattered image is 63.91 × 41.55 μm^2 and that the size corresponding to each 1 mm movement of the colorimeter is 0.1522 μm. In each direction, every time we moved the colorimeter by 1 mm, we recorded the *L* value once and considered the arithmetic average of the thickness of the compatible layer obtained from each direction as the final experimental result.

According to the principle described above, we tested the thickness of the compatible layer on the backscattered electron images of samples 1–5. At the same time, to test the effect of the silane coupling agent KH-172 on the formula, backscattered electron images were also taken for the control group without KH-172, and the thickness of the compatibility layer was tested. The results are shown in Figure 7. From Figure 7, we find that the use of the vulcanizing agent has little effect on the thickness of the compatibility layer between the two phases. At the same time, we find that KH-172 can effectively increase the thickness of the compatibility layer between the two phases. After calculation, the thickness of each group of compatible layers can be increased by nearly 35% after adding KH-172. In addition, we also tested the mechanical properties of the samples with and without KH-172 added, and the test results are shown in Figures 8 and 9. From Figure 8, we see that the addition of KH-172 can effectively increase the tensile strength and elongation at the break of the material. It should be noted that the tensile permanent deformation is an index to measure the elasticity. The smaller the value itself, the better the resilience performance of the material. Therefore, the reduction of the tensile permanent deformation of each group of materials after adding KH-172 means the improvement of the elastic properties of the material. After calculation, compared with the mechanical data without KH-172, the tensile strength is increased by about 25%, the elongation at the break is increased by about 15%, and the tensile permanent deformation is improved by about 20%. Combined with the characterization results of the size of the compatible layer in Figure 7, we believe that the improvement of the tensile and elastic properties of this modified TPV blend system by KH-172 is due to the increase in the size of the compatible layer, that is, KH-172 can increase the interaction between the two phases in the TPV system to a certain extent. From Figures 8 and 9, in addition to the effect of KH-172, comparing the mechanical property data of each group with the different vulcanizing agent content, we found that when the vulcanizing agent content was 1 phr, the tensile strength was the largest, the tensile permanent deformation was the smallest, and the elongation at the break decreased gradually with the increase of the content of the vulcanizing agent. The decrease in elongation at the break was mainly related to the increase in the degree of the internal cross-linking of the polymer. As the cross-linked structure becomes denser, the degree of stretching of the polymer chain would decrease, and this phenomenon occurred. However, the tensile strength and elastic properties first improved with the increase of the vulcanizing agent content, and then the properties decreased greatly, both of which were mainly related to the degree of crosslinking and the size of the dispersed phase. When the amount of vulcanizing agent was increased from 0 phr to 1 phr, both the PSR and SBS were cross-linked to form a three-dimensional network structure. At the same time, it can be seen from Figure 5A–C that the size of the dispersed phase was significantly reduced, which made the material show good elasticity of PSR and good tensile properties of SBS-SEBS. However, as the amount of vulcanizing agent continued to increase, both the PSR and SBS demonstrated over-vulcanization. At the same time, it was difficult to make the shear force of the rotor disperse the dispersed phase well, resulting in a significant increase in the size of the dispersed phase. This can be seen from Figure 5D,E, where the overall tensile property and elasticity of the material declined significantly.

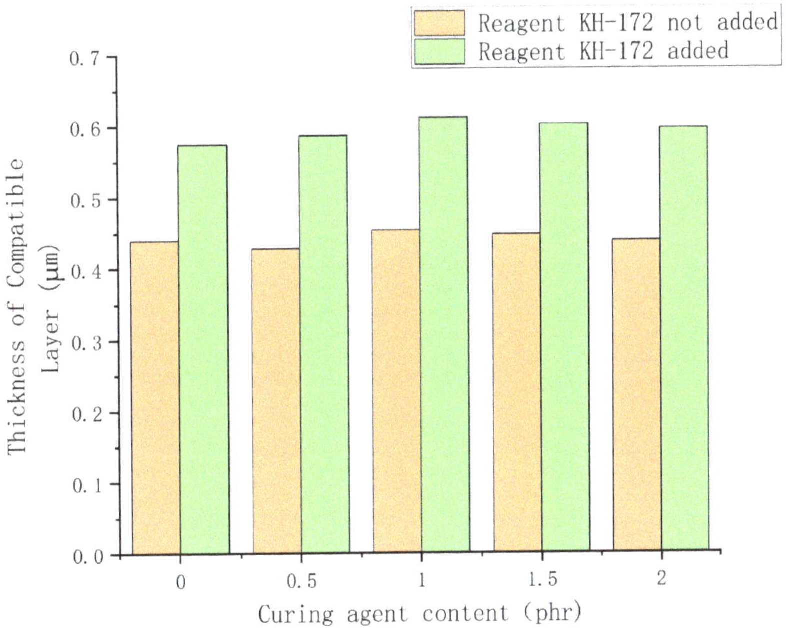

Figure 7. Thickness histogram for quantitative characterization of the compatible layer.

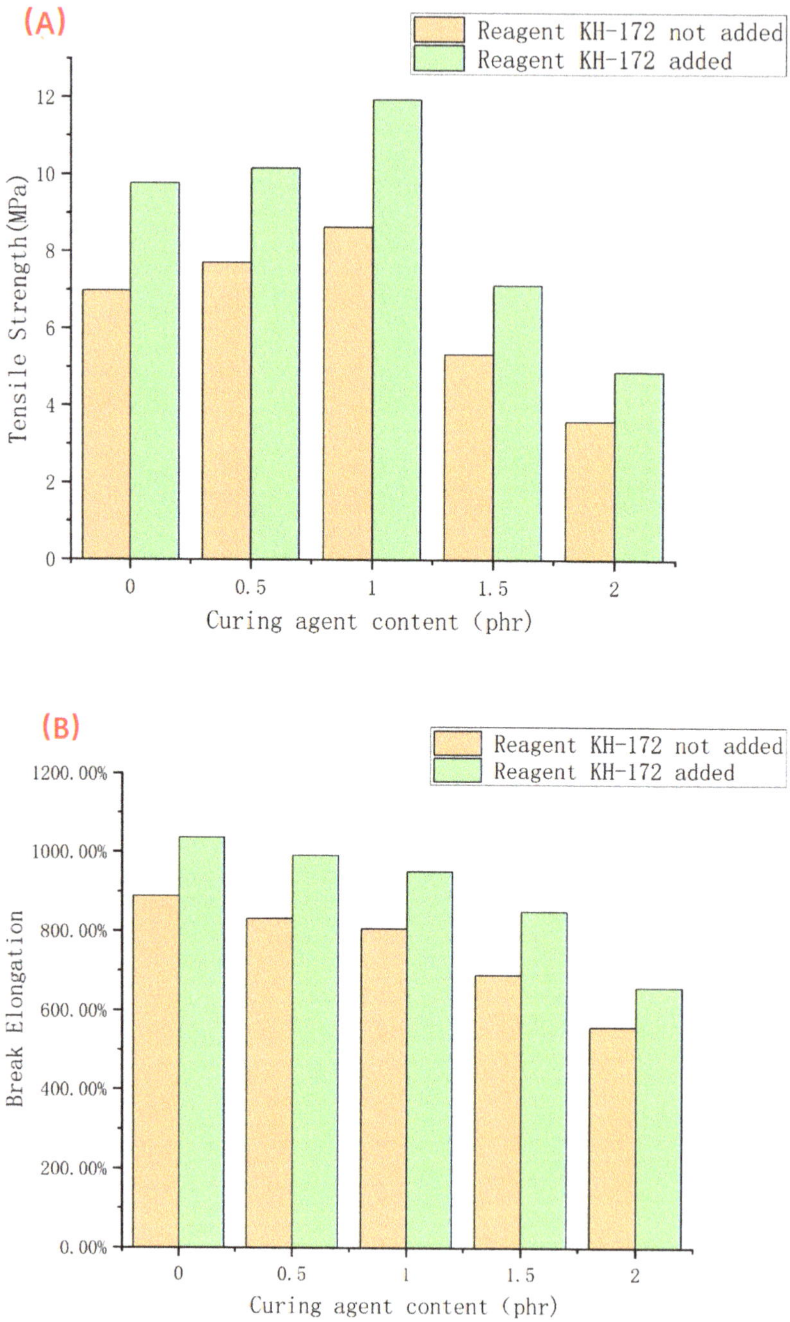

Figure 8. Histogram of tensile strength and elongation at the break of samples 1–5 ((**A**) is the change in tensile strength of each group of samples before and after adding KH-172, and (**B**) is the change of elongation at break of each group of samples before and after adding KH-172.).

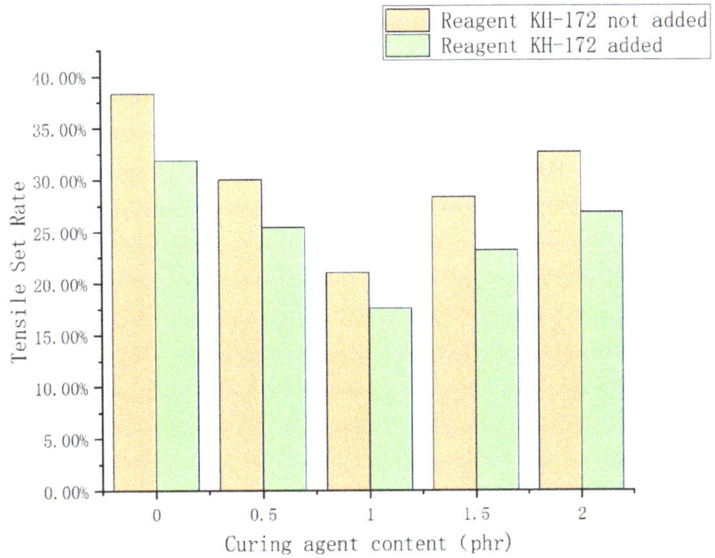

Figure 9. Histogram of tensile permanent deformation of samples 1–5.

3.4. Analysis of the Experimental Results of Dynamic Mechanical Test and Low-Temperature Hardness

As shown in Figure 10A, a graph of the loss tangent value of samples 1–5, pure PSR and SEBS-SBS blends containing 50% of each. From the figure, we see that the blend of SEBS-SBS has two peaks, corresponding to the temperatures of −90 °C and 102 °C, respectively. These two peaks represent the polybutadiene segment and the polystyrene segment, respectively. The pure PSR peaks at −48 °C reflect the glass transition temperature of PSR. There are three peaks in the loss tangent curve corresponding to samples 1–5. Because both PSR and SBS exist in the sample, the peak of the glass transition temperature of PSR and the peak of the polybutadiene section of SBS appear. Although there is a certain degree of overlap, the peak temperatures of −75 °C, −55 °C, and 94 °C can also be distinguished. Compared with the pure components in the sample, the peaks of the polybutadiene segment of SBS are shifted by nearly 15 °C, the peaks of the polystyrene segment are shifted by nearly 8 °C, and the peaks of PSR are shifted by nearly 7 °C. Especially in the low-temperature region, the peaks of the two components are close, which indicates that the two components in the modified TPV system have a certain degree of compatibility.

Figure 10B shows the storage modulus curves of samples 1–5. To facilitate the observation of the changes in the magnitude of the storage modulus, we took the logarithm of the storage modulus value. The storage modulus is usually used to measure the elasticity of elastomers. People often regard the storage modulus as an index of the ability of materials to store elastic deformation energy. This concept is simply the ratio of stress to elastic deformation. According to the description of the definition, we find that under the same stress, the larger the storage modulus, the smaller the elastic deformation. If this phenomenon occurs at low temperatures, it means that the resilience of the rubber becomes poor. This can limit the use of the material at a certain temperature. From Figure 10B, we see that when the temperature is higher than −20 °C, the storage modulus of the 5 groups of samples is less than 10 MPa. However, the value of the storage modulus increased by an order of magnitude when the temperature dropped to −60 °C. The storage modulus increases by another order of magnitude when the temperature continues to drop to −80 °C. In addition, we compared the storage modulus curves of samples 1–5 with those of pure PSR and unvulcanized SBS/SEBS blends and found that the storage modulus of silicone rubber is less than 100 MPa in the entire temperature range. This indicates that PSR has good low-temperature resistance, that is, it can maintain good elasticity at low temperatures. The storage modulus of the unvulcanized SBS/SEBS blends increases sharply at around −40 °C, and the storage modulus exceeds 1000 MPa when the temperature reaches nearly −65 °C. Combined with the above analysis of the magnitude change of the storage modulus of samples 1–5, we also see that the addition of PSR can significantly improve the low-temperature performance of SBS/SEBS.

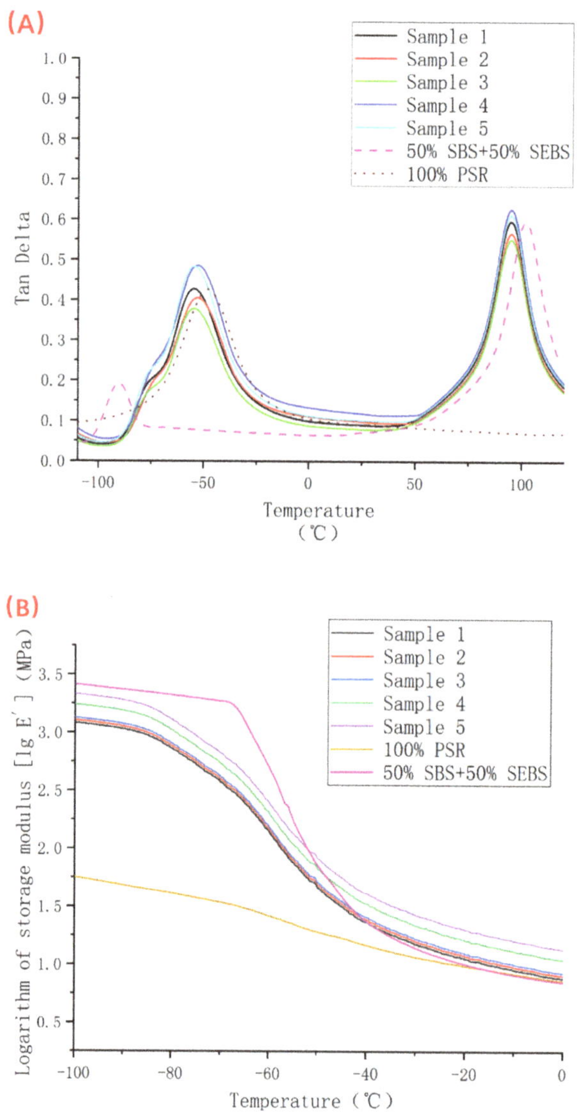

Figure 10. Dynamic mechanical test results of samples 1–5 (where (**A**) is the tangent value curve of loss angle from −110 °C to 120 °C and (**B**) is the logarithm of the storage modulus from −100 °C to 0 °C).

We counted the storage modulus values of each group at −60 °C, as shown in Figure 11A, and also tested the hardness changes at room temperature and −60 °C, as shown in Figure 11B. From Figure 11B, we see that when the amount of the vulcanizing agent is more than 1 phr, the hardness of the material will increase sharply; in other words, the elasticity of the material will have significantly deteriorated. Combined with the data in Figure 11A, we found that the storage moduli of samples 4 and 5 with significantly increased hardness were both greater than 200 MPa. This means that in the temperature range where the storage modulus is greater than this value, the material will lose elasticity to a certain extent. At the same time, we also easily find from Figure 11A that with the addition of too much vulcanizing agent, the storage modulus of the material at low temperature will increase significantly, that is, the low-temperature performance will decrease. The degree of crosslinking is too deep, thus the importance of proper control of the amount of vulcanizing agent can be seen.

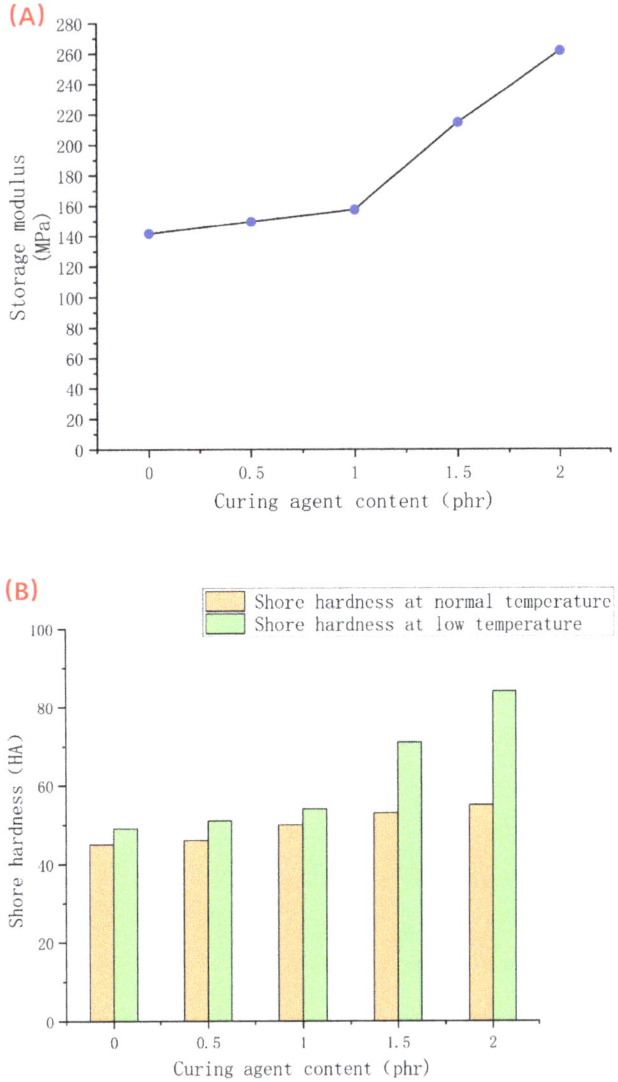

Figure 11. Low-temperature hardness change of samples 1–5 at −60 °C and broken line statistic diagram of the storage modulus at −60 °C ((**A**) is the broken line statistic diagram of the storage modulus at −60 °C and (**B**) is the column diagram of low-temperature hardness change at −60 °C).

Furthermore, as shown in Figure 12, we plotted the loss factor versus the logarithm value of the storage modulus E′ at different frequencies (frequency range 1–20 Hz) at −60 °C. In this way, we observe the dependence of the response time and the temperature of the material under dynamic stress well [38,39]. From the figure, we see that there is a correlation between the loss factor and the storage modulus at different frequencies for each group of samples at −60 °C. Moreover, this correlation persists regardless of the vulcanization content, which is directly related to the degree of cross-linking. In addition, the storage modulus of each group of samples increases with the increase in frequency. According to the time-temperature equivalent principle, increasing the frequency has the same effect as lowering the temperature or reducing the force time. Therefore, the rigidity of the materials is increased, that is, the storage modulus is increased. It also provides a reference for the selection of the suitable application environments of materials.

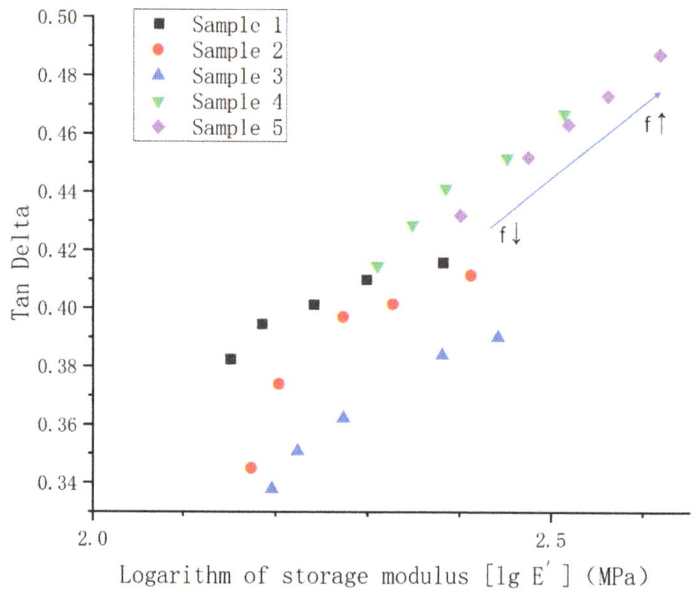

Figure 12. Relationship between the loss factor and storage modulus of samples 1–5 at −60 °C.

3.5. Analysis of Melt Index Test Results

Figure 13 is a bar graph of the melt index of samples 1–5 tested under a load weight of 5 kg. The figure shows that the melt index can reach 19.66 g/10 min when no vulcanizing agent is added, which demonstrates that the modified TPV system has good fluidity during simple blending and is relatively easy to be extruded during processing. When the vulcanizing agent content increased from 0 phr to 1 phr, the melt index decreased by 32.81% and 25.36%, respectively. The main reason for this phenomenon is that with the increase of the vulcanizing agent content, the degree of cross-linking inside the two phases of the blend increases continuously, and the three-dimensional grid structure becomes more and more compact, which greatly reduces the fluidity of the blend. Then, when the vulcanizing agent content continued to increase to 2 phr, the melt index decreased by 74.85% and 100%, respectively. That is, when the vulcanizing agent content is 2 phr, the blend cannot be extruded from the melt flow rate tester. Combining Figure 5D,E, we see that the size of the PSR as a dispersed phase is very large, and the degree of crosslinking is too large, which makes it difficult for the PSR phase to be divided into smaller sizes by the shear force of the rotor, which also makes it difficult for the melt blend containing the PSR phase with the larger size to be extruded, or even impossible to be extruded. The melt index is an important indicator for determining processing technology. The data we tested here can provide a certain reference for the process selection of the products made from this blend system.

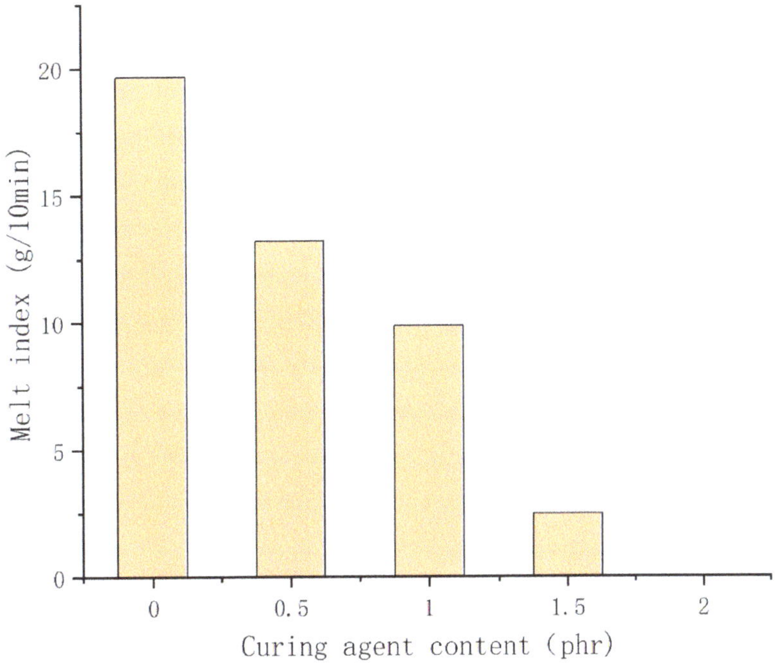

Figure 13. Histogram of the melt index of samples 1–5.

4. Conclusions

In summary, we successfully prepared a modified TPV using PSR/SEBS/SBS, which exhibited good mechanical properties and low-temperature properties. For the formulation of this material, we deeply studied the effect of the amount of vulcanizing agent on the material properties and microscopic phase distribution. It was found that when the dosage of the vulcanizing agent was 1 phr, the properties of the material were the best, the size of the dispersed phase was the smallest, and the dispersion of the dispersed phase was the best. In addition, through the analysis of the backscattered images and the test of the mechanical properties, we found that the silane coupling agent KH-172 can also significantly improve the tensile properties and elastic properties of this TPV system. Finally, we tested the thermal fluidity of the material and found that when the vulcanizing agent was used in excess of 1 phr, the fluidity of the melt was greatly reduced. In general, the new TPV we developed and prepared is expected to expand the application scenarios of silicone rubber products, and also provide a new idea for the development of similar products.

Author Contributions: Methodology, X.C. and B.H.; project administration, X.C. and B.H.; resources, X.C. and B.H.; supervision, X.C. and B.H.; conceptualization, X.C.; writing—original draft, C.Z.; validation, C.Z.; writing—review & editing, C.Z.; visualization, C.Z.; data curation, C.Z.; investigation, C.Z.; formal analysis, C.Z. All authors have read and agreed to the published version of the manuscript.

Funding: This research received no external funding.

Institutional Review Board Statement: Not applicable.

Data Availability Statement: The data presented in this study are available on request from the corresponding author.

Conflicts of Interest: The authors declare no conflict of interest.

References

1. Shit, S.C.; Shah, P. A review on silicone rubber. *Natl. Acad. Sci. Lett. India* **2013**, *36*, 355–365. [CrossRef]
2. Deshpande, G.; Rezac, M.E. The effect of phenyl content on the degradation of poly(dimethyl diphenyl) siloxane copolymers. *Polym. Degrad. Stab.* **2001**, *74*, 363–370. [CrossRef]
3. Lee, J.H.; Bae, J.W.; Choi, M.C. Effect of phenyl vinyl methyl silicone (PVMQ) on low temperature sealing performance of fluorosilicone composites. *Elastomers Compos.* **2021**, *56*, 209–216.
4. Huang, Y.H.; Mu, Q.H.; Su, Z.T. High and Low Temperature Resistance of Phenyl silicone rubber. *IOP Conf. Ser. Mater. Sci. Eng.* **2021**, *1048*, 012001. [CrossRef]
5. Aksut, D.; Sen, M.; Demeter, M. Effect of radiation on vinyl-methyl-polysiloxane and phenyl-vinyl-methyl-polysiloxane elastomers cured with different co-agents: Comparative study of mechanical and relaxation properties. *Radiat. Phys. Chem.* **2019**, *158*, 87–93. [CrossRef]
6. Zhang, Y.J.; Fan, M.S.; Li, X.P. Silicon-containing functionalized polyolefin: Synthesis and application. *Prog. Chem.* **2020**, *32*, 84–92.
7. Shi, C.R.; Zong, C.Z.; Pan, H.W. A method to essentially increase the compatibility of thermoplastic polyurethane/methyl vinyl silicone rubber based thermoplastic vulcanizate. *J. Appl. Polym. Sci.* **2022**, *139*, e52219. [CrossRef]
8. Zhao, G.J.; Tian, K.; Pan, Q.; Zhang, Q.; Deng, H.; Fu, Q. A Novel Method to Incorporate Functional Filler into TPSiV for Balanced Physical Properties. *Compos. Sci. Technol.* **2021**, *213*, 108925. [CrossRef]
9. Radusch, H.J. Dynamic vulcanizates on the basis of polypropylene/rubber mixtures. *Kautsch Gummi Kunstst* **1990**, *43*, 767–769.
10. School, R. TPE production and use in the 1990s. *Rubber World* **1990**, *201*, 15.
11. Choudhury, N.R.; Bhowmick, A.K. Strength of thermoplastic elastomers from rubber-polyolefin blends. *J. Mater. Sci.* **1990**, *25*, 161–167. [CrossRef]
12. Radusch, H.J. Dynamic vulcanization of EPDM-polypropylene mixtures. *Kautsch Gummi Kunstst* **1991**, *44*, 1125–1128.
13. Islam, S.K.S.; Singh, S.K.; Ransinchung, R.N.G.D. Performance Deterioration of SBS-Modified Asphalt Mix: Impact of Elevated Storage Temperature and SBS Concentration of Modified Binder. *J. Mater. Civ. Eng.* **2022**, *34*, 04021475. [CrossRef]
14. Kok, B.V.; Erkus, Y.; Yilmaz, M. Effects of triple modification on the mechanical properties and the cost of asphalt mixtures. *Environ. Eng. Manag. J.* **2022**, *21*, 603–613. [CrossRef]
15. Murthe, S.S.; Sreekantan, S. Study on the Physical, Thermal and Mechanical Properties of SEBS/PP (Styrene-Ethylene-Butylene-Styrene/Polypropylene) Blend as a Medical Fluid Bag. *Polymers* **2022**, *14*, 3267. [CrossRef] [PubMed]
16. Gao, W.; Yu, B.; Li, S.; Chen, S.; Zhu, Y.; Zhang, B.; Zhang, Y.; Cai, H.; Han, B. Preparation and properties of reinforced SEBS-based thermoplastic elastomers modified by PA6. *Polym. Eng. Sci.* **2022**, *62*, 1052–1060. [CrossRef]
17. Zhao, H.; Gao, W.-C.; Li, Q.; Khan, M.R.; Hu, G.-H.; Liu, Y.; Wu, W.; Huang, C.-X.; Li, R.K. Recent advances in superhydrophobic polyurethane: Preparations and applications. *Adv. Colloid Interface Sci.* **2022**, *303*, 102644. [CrossRef]
18. He, C.; Li, B.; Ren, Y.; Lu, W.; Zeng, Y.; He, W.; Feng, A. How the Crosslinking Agent Influences the Thermal Stability of RTV Phenyl Silicone Rubber. *Materials* **2019**, *12*, 88. [CrossRef]
19. Shiva, M.; Dallakeh, M.K.; Ahmadi, M.; Lakhi, M. Effects of silicon carbide as a heat conductive filler in butyl rubber for bladder tire curing applications. *Mater. Today Commun.* **2021**, *29*, 102773. [CrossRef]
20. Nurhajati, D.W.; Lestari, U.R.; Setyorini, I.; Mayasari, H.E. Effect of carbon black/Bi(2)O(3)ratio on the vulcanization characteristic and mechanical properties of rubber compound. *Macromol. Symp.* **2020**, *391*, 1900171. [CrossRef]
21. Awasthi, P.; Banerjee, S.S. Design of ultrastretchable and super-elastic tailorable hydrophilic thermoplastic elastomeric materials. *Polymer* **2022**, *252*, 124914. [CrossRef]
22. Shahroodi, Z.; Katbab, A.A. Preparation and characterization of peroxide-based dynamically vulcanized thermoplastic elastomer of poly (lactic acid)/chloroprene rubber. *Polym. Eng. Sci.* **2022**, *62*, 1485–1495. [CrossRef]
23. Kana, M.; Saito, H. Tensile properties and interfacial adhesion of silicone rubber/polyethylene blends by reactive blending. *J. Appl. Polym. Sci.* **2018**, *135*, 46192.
24. Yao, N.Q.; Wang, H.B.; Zhang, L.Q. One-pot Hydrothermal Synthesis of Silane-Functionalized Carbon Nanodots as Compatibilizers for the Immiscible TPU/MVQ blends. *Appl. Surf. Sci.* **2020**, *530*, 147124. [CrossRef]
25. Toshio, I. Selective crosslinking in polymer blends I. Novel selective crosslink systems for polypropylene/unsaturated elastomer blends. *J. Appl. Polym. Sci.* **1994**, *54*, 709.
26. Chambers, S.A. Elastic-scattering and interference of backscattered primary, auger and X-ray photoelectrons at high kinetic-energy-principles and applications. *Surf. Sci. Rep.* **1992**, *16*, 261–331. [CrossRef]
27. Autrata, R. Backscattered electron imaging using single-crystal scintillator detectors. *Scanning Microsc.* **1989**, *3*, 739–763.
28. Yoo, H.-S.; Kim, Y.-H.; Son, H.-T. The Effects of TiC Content on Microstructure of Modified A6013-3wt.%Si Alloy Powder Compact. *J. Korean Powder Met. Inst.* **2022**, *29*, 28–33. [CrossRef]
29. Brackmann, L.; Schuppener, J.; Röttger, A.; Weber, S. Influence of a Partial Substitution of Co by Fe on the Phase Stability and Fatigue Behavior of a CoCrWC Hard Alloy at Room Temperature. *Met. Mater. Trans. A* **2022**, *53*, 2708–2723. [CrossRef]
30. Rößler, C.; Zimmer, D.; Trimby, P.; Ludwig, H.-M. Chemical–crystallographic characterisation of cement clinkers by EBSD-EDS analysis in the SEM. *Cem. Concr. Res.* **2022**, *154*, 106721. [CrossRef]

31. Chen, X.; Wang, G.; Dong, Q.; Zhao, X.; Wang, Y. Microscopic characterizations of pervious concrete using recycled Steel Slag Aggregate. *J. Clean. Prod.* **2020**, *254*, 120149. [CrossRef]
32. Goizueta, G.; Chiba, T.; Inoue, T. Phase morphology of polymer blends: Scanning electron microscope observation by backscattering from a microtomed and stained surface. *Polymer* **1992**, *33*, 886. [CrossRef]
33. Goizueta, G.; Chiba, T.; Inoue, T. Phase morphology of polymer blends: 2. SEM observation by secondary and backscattered electrons from microtomed and stained surface. *Polymer* **1993**, *34*, 253–256. [CrossRef]
34. Masters, R.C.; Wan, Q.; Zhang, Y.; Dapor, M.; Sandu, A.M.; Jiao, C.; Zhou, Y.; Zhang, H.; Lidzey, D.G.; Rodenburg, C. Novel organic photovoltaic polymer blends: A rapid, 3-dimensional morphology analysis using backscattered electron imaging in the scanning electron microscope. *Sol. Energy Mater. Sol. Cells* **2017**, *160*, 182–192. [CrossRef]
35. Wang, W.; Huang, J.; Gong, Z.; Fan, J.; Cao, L.; Chen, Y. Biobased PLA/NR-PMMA TPV with balanced stiffness-toughness: In-situ interfacial compatibilization, performance and toughening model. *Polym. Test.* **2019**, *81*, 106268. [CrossRef]
36. Khanra, S.; Sreenivasan, P.; Das, S.; Hore, R.; Ganguly, D.; Chattopadhyay, S. Immobilization of a biobased process aid at the interface for binary silicone and fluoroelastomer based super specialty blends with silica for enhanced compatibility. *J. Mater. Sci.* **2022**, *57*, 13974–13990. [CrossRef]
37. Goldstein, J.I.; Newbury, D.E.; Michael, J.R. *Scanning Electron Microscopy and X-ray Microanalysis*; Springer: Berlin/Heidelberg, Germany, 2017.
38. Çakmak, U.; Fischlschweiger, M.; Graz, I.; Major, Z. Adherence Kinetics of a PDMS Gripper with Inherent Surface Tackiness. *Polymers* **2020**, *12*, 2440. [CrossRef]
39. Tschoegl, N.W.; Knauss, W.G.; Emri, I. The Effect of Temperature and Pressure on the Mechanical Properties of Thermo- and/or Piezorheologically Simple Polymeric Materials in Thermodynamic Equilibrium—A Critical Review. *Mech. Time Depend. Mater.* **2002**, *6*, 53–99. [CrossRef]

polymers

MDPI

Article

Molecular Dynamics Simulation of the Thermomechanical and Tribological Properties of Graphene-Reinforced Natural Rubber Nanocomposites

Zepeng Wang *, Minglong Su, Xinwu Duan, Xiulong Yao, Xiaoying Han, Junping Song and Lianxiang Ma

College of Electromechanical Engineering, Qingdao University of Science and Technology, Qingdao 266061, China
* Correspondence: 02546@qust.edu.cn; Tel.: +86-139-5325-8705

Abstract: The thermomechanical and tribological properties of graphene (GNS)-reinforced NR were investigated using molecular dynamics (MD) simulations. The amorphous molecular dynamics models of two nanocomposites, i.e., natural rubber (pure NR) and graphene/natural rubber (GNS/NR), were established. In addition, the thermodynamic properties of the two materials, before and after the incorporation of graphene into the natural rubber matrix, were investigated through analytical comparison. The results showed that after the graphene was added to the rubber matrix as a reinforcing material, the elastic modulus and shear modulus were increased by 110% and 94.8%, respectively, the tensile property was increased by 178%, the overall thermal conductivity of the composite system was increased by 59%, the glass transition temperature increased from 223 K to 236 K, and the rigidity of the material matrix was significantly improved. The inherent interactions and wear mechanisms of the polymer nanocomposites were discussed at the atomic scale by analyzing the changes in temperature, atomic velocity, relative atomic concentration, and radial distribution functions at the friction interface in the thickness direction.

Keywords: molecular dynamics simulation; nanocomposites; thermodynamic properties; tribological properties

Citation: Wang, Z.; Su, M.; Duan, X.; Yao, X.; Han, X.; Song, J.; Ma, L. Molecular Dynamics Simulation of the Thermomechanical and Tribological Properties of Graphene-Reinforced Natural Rubber Nanocomposites. *Polymers* **2022**, *14*, 5056. https://doi.org/10.3390/polym14235056

Academic Editor: Alexey V. Lyulin

Received: 26 September 2022
Accepted: 15 November 2022
Published: 22 November 2022

1. Introduction

Polymer composites have become one of the most promising materials in the field of materials science today due to their excellent mechanical, thermal, electrical, and chemical properties [1,2], and they are widely used in production and industrial engineering [3]. It is therefore necessary to deeply understand their ontological properties in terms of their physical microenvironment and chemical structure [4,5], especially their thermomechanical and tribological properties.

With the rapid development of polymer nanocomposites, molecular dynamics (MD) simulation has become an effective alternative method for studying and predicting material properties at the microscopic atomic scale, providing microscopic information and mechanisms of molecular interactions. Simultaneously, MD simulation can also be used to analyze the experimental results. Therefore, molecular dynamics simulation studies are essential to evaluate and study the thermodynamic properties of polymer nanocomposites. M. Ding et al. [6] calculated the thermomechanical properties of cross-linked epoxy resin/functionalized carbon nanotube composites based on molecular dynamics simulations, and the simulation results showed that the overall thermal conductivity of epoxy nanocomposites doped with carbon nanotubes were all improved, and the doping of carbon nanotubes led to the improvement in the glass transition temperature and mechanical properties of the composite system. Y. Li et al. [7–11] investigated the mechanical and frictional wear properties of carbon nanotube/NBR (nitrile-butadiene rubber) composites using molecular dynamics simulations. The results showed that the introduction of CNTs reduced the friction coefficient of the composites by about 38% and the average wear rate by about 60% under different normal loads. In addition, the tribological enhancement of CNTs/polymer composites was further investigated. Q, Xue et al. [12] used CNTs as a reinforcing agent and demonstrated that they could inhibit the adhesion and wear of the PI matrix, thus reducing the friction coefficient and frictional wear volume. H. Chen et al. [13] simulated the CNT pull-out process and a series of friction processes between the composite and metal substrate by MD simulation to explain the motion state between the atoms of each part from the energy perspective, studying the force on the CNT and the change in the displacement of each part to infer the motion attitude. K Pan et al. [14] simulated the detachment process of cross-linked polydimethylsiloxane (PDMS) from functionalized graphene (fG) using the molecular dynamics simulation method using ReaxFF potential function. The effects of graphene modified by three different functional groups—hydroxyl (-OH), amino (-NH2), and carbon groups (-CH3)—on the interfacial mechanical properties of the composites were investigated. The results showed that different functional groups had a significant influence on the interfacial mechanical properties of the composites.

In this study, the effects of hydrogenated graphene on the thermomechanical and tribological properties of natural rubber were investigated. Firstly, a pure natural rubber (polyisoprene pure rubber) model and hydrogenated graphene/natural rubber (GNS/NR) composite model were constructed using Materials Studio [15–18] software to simulate and calculate the tensile properties, shear modulus, glass transition temperature, and thermal conductivity of the materials. Next, a molecular model of the metallic iron atomic layer and the polymer matrix was developed to simulate the frictional wear behavior. Finally, by analyzing the microscopic information, such as atomic concentration, atomic velocity distribution, and radial distribution function at the friction interface, the microscopic mechanism of the thermodynamic and frictional wear properties of graphene-enhanced natural rubber was explored and revealed at the microscopic level. This study overcomes the limitations of previous frictional wear macroscopic tests and provides a new method and theoretical guidance for the study of inorganic nanomaterial-reinforced polymer composites.

2. Materials and Methods

2.1. Model Construction

In this study, a nanocomposite model system with periodic boundary conditions was developed using Materials Studio (8.0). The "Amorphous Cell Package" and "Forcite" modules in Materials Studio were used for modeling and simulation. The COMPASS II force field [19], which can provide interatomic interactions and qualitatively describe the physical and mechanical properties [20], as well as the thermodynamic properties of the polymer, was used in the simulations. The first step was to establish a periodic cubic lattice with the dimensions of 45 Å × 45 Å × 45 Å and to construct a molecular chain of polyisoprene $(C5H8)_n$ natural rubber with a repeating unit of a polymerization degree of 30. Then, a monolayer graphene GNS with the dimensions of 11.30 Å × 12.30 Å was constructed, in which the edge of GNS was functionalized with hydrogen atoms, and subsequently, the hydrogenated monolayer graphene was placed at the center of the lattice, as shown in Figure 1.

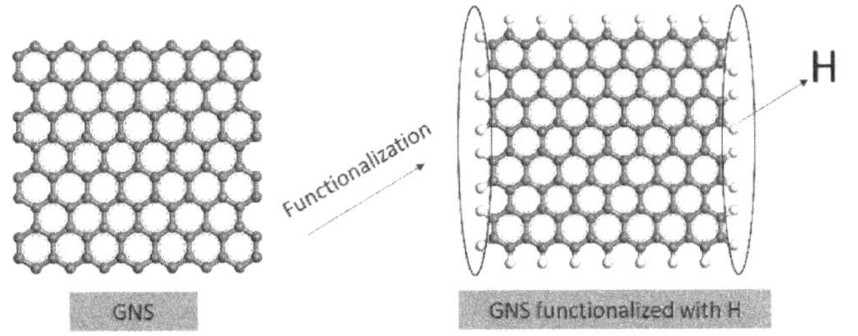

Figure 1. Graphene functionalized with hydrogen atoms.

Eventually, the polyisoprene molecular chains were gradually and randomly filled into the lattice, according to Monte Carlo rules, until the lattice density reached the theoretical value of 0.92 g/cm^3. For the Pure NR amorphous molecular model, the molecular model was constructed in the same way, with a preset density of 0.92 g/cm^3. The resulting model is shown in Figure 2.

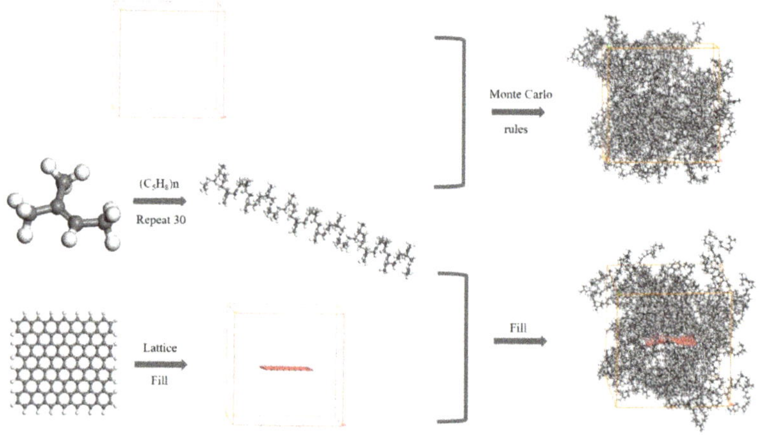

Figure 2. Molecular model: NR amorphous molecular model and GNS/NR amorphous molecular model.

2.2. Model Optimization

Since the internal energy of the constructed molecular models was too high and the molecules were in a highly unstable state, the conjugate gradient method was used to calculate the total potential energy of the two amorphous molecular models in a geometrically optimized manner to achieve the minimal local energy configuration of the molecular system. The energy convergence accuracy reached the root mean square value of 0.00001 kcal/mol, and the force convergence accuracy reached 0.0001 kcal/mol/Å to obtain the global minimum energy configuration. In order to improve the accuracy of the simulation result, the model was further equilibrated by performing five annealing cycles of simulation in the temperature range of 150 K to 350 K. It was required to obtain the molecular dynamics equilibrium of the two molecular conformation systems at the reference temperature of 150 K, to set the temperature growth step to 50 K, and to run a constant temperature isotropic (NVT system synthesis) system for 50 ps at the target temperature. After the simulation of several annealing cycles at high and low temperatures, the structure of the molecular system was further relaxed, and the model configuration was gradually rationalized to a stable state with minimal local energy. Finally, the two systems were subjected to an isothermal isobaric (NPT system synthesis) process of molecular dynamics equilibrium at room temperature 298 K, pressure 101 KPa, and time step 1 ps. The algorithm of Nose and Berendsen was used to control the temperature and pressure during the simulation. When the energy and density of the entire system do not change significantly with time, we believe that the molecular model has reached the most stable state, with minimal energy. The final densities of the pure NR model and the GNS/NR model were 0.897 g/cm^3 and 0.959 g/cm^3, respectively, which were consistent with the true density of natural rubber polyisoprene of 0.92 g/cm^3.

2.3. Thermodynamic and Tribological Simulation

The mechanical properties of both the pure NR and GNS/NR composites were calculated by the method of constant strain technique to calculate the stress–strain properties and shear modulus of the materials. The glass transition temperature is determined by calculating the specific volumes of the two materials at different temperatures, obtained by linearly fitting the resulting data and comparing them. Based on Fourier's law, non-equilibrium molecular dynamics simulations were used to calculate the thermal conductivity of the two materials separately.

In order to simulate the tribological properties of the pure NR and GNS/NR composites, two three-layer tribological sub-models were established. Combined with the actual working conditions of rubber tires, the iron atom layer was selected for the upper and lower layers of the friction sub-model, and the iron atom layer size was 45 Å × 45 Å × 11.5 Å. Pure NR and GNS/NR were used as intermediate layers to fix the upper and lower iron atomic layers, and the two friction sub-models were optimized separately using the above-mentioned method of optimizing the amorphous model, and the fixation of the metal atomic layer was removed after the optimization was completed. To obtain the tribological properties of the polymer composite layer and the metal layer, a positive pressure of 1.01×10^{-4} GPa and a relative sliding friction velocity of 0.1 Å/ps were given to the upper iron atomic layer, the temperature was set to 298 K in an ambient environment, and the whole system was run for 600 ps under the NVT system synthesis to obtain the trajectory file of the frictional wear situation over time, which was used to analyze the material frictional wear properties.

3. Results and Discussions

3.1. Thermodynamic Property

3.1.1. Tensile Properties Simulation

The tensile properties of the pure natural rubber and graphene/natural rubber nanocomposites were simulated using the constant strain method [21,22]. The tensile properties of the pure NR and GNS/NR composites can be calculated through the uniaxial stretching process in the X-direction to obtain the stress–strain curves for both materials, as shown in Figure 3.

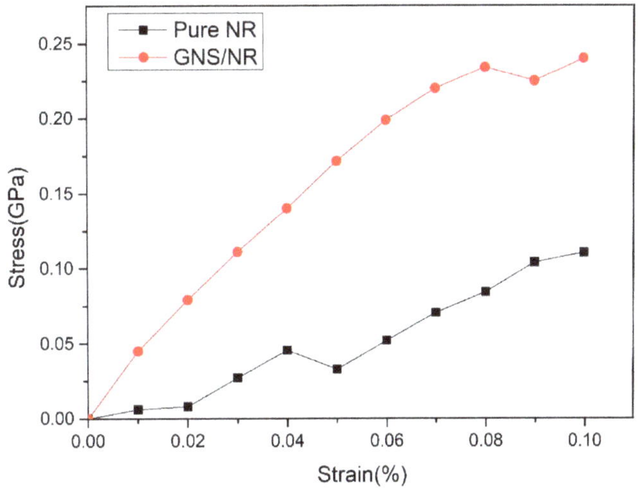

Figure 3. Stress-strain curves of the pure NR and GNS/NR composites.

It can be seen from Figure 3 that in the strain range of 0 to 0.08, the tensile stress of the matrix of the two material showed an increasing trend, and the tensile stress of the GNS/NR composite matrix increased to 0.24 GPa in a linear manner. By comparing these results with literature data [23], this stress–strain behavior was within a reasonable range of variation, and the accuracy of the calculation method for the tensile behavior using molecular dynamics simulations was verified.

The points with large errors were eliminated, and the slope of the fitted stress–strain curve was used to calculate the modulus of elasticity. The modulus of elasticity of the two material substrates, pure NR and GNS/NR, was 1.150 GPa and 2.416 GPa, respectively. When hydrogenated GNS was added to the material as a reinforcement system, the modulus of elasticity of the NR substrate was increased by 110%, and the tensile properties were improved by 178% (at the strain of 0.08). This phenomenon is due to the adsorption of VDW and electrostatic forces on the surface of GNS. A spatially reinforcing interface was formed between GNS and the natural rubber nanopolymer matrix, which made the polymer chains adsorbed tightly around the GNS. When the whole system was subjected to external forces, the adsorption of GNS on the polymer matrix limited the shedding of polymer chains within the matrix. Consequently, when the GNS/NR material matrix was subjected to an external stretching force, a larger stretching force was needed to achieve the same strain; thus, the material matrix had an increased elastic modulus. The high elastic modulus prevented cracking during stretching and further improved the tear strength of the matrix, thus extending the service life of the rubber composite matrix.

In order to explore this reaction more deeply, the mechanism was verified and analyzed. The radial distribution function between GNS and the natural rubber polymer matrix during the stretching process was calculated and extracted for analysis. As shown in Figure 4, after adding GNS to the rubber material matrix as a reinforcing material, when the contact distance between atoms in the Z direction was greater than 10 Å during the stretching process, the RDF values of the radial distribution function between the GNS and the atoms within the polymer matrix tended to be in dynamic equilibrium. The unique two-dimensional lamellar structure of GNS, with a higher specific surface area, could adsorb and adhere to more of the polymer matrix. When the composite matrix was subjected to tensile forces, the VDW forces, as well as the electrostatic adsorption between the upper and lower surfaces of the GNS and the polymer matrix, prevented the polymer chains from moving in the tensile direction. Therefore, higher tensile stress is required to obtain the same tensile strain as that of pure NR, which shows the better mechanical properties of the composite.

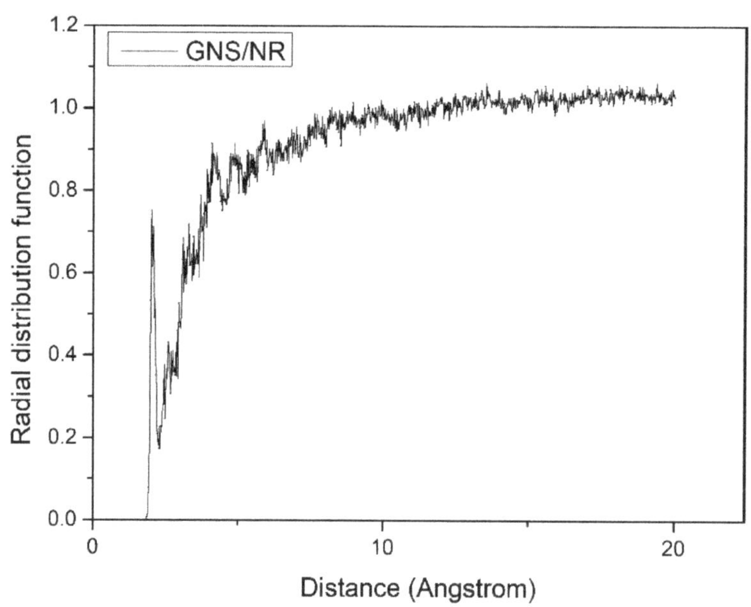

Figure 4. Radial distribution function between graphene and the natural rubber matrix during the tensile process.

3.1.2. Shear Modulus and Bulk Modulus

To further investigate the effect of graphene on the mechanical properties of rubber nanomaterials, the bulk modulus (K) and the shear modulus (G) of both the pure NR and GNS/NR were calculated based on the optimized final structure. The constant strain approach was employed by exerting 20 in 6 directions (i.e., x, y, z, yz, xz, xy) of the systems. Finally, the 6*6 stiffness matrices (C_{ij}) and the compliance matrices (S_{ij}) were obtained using the following equations:

$$C_{ij} = \sigma_i \tag{1}$$

$$S_{ij} = \varepsilon_j / \sigma_i \tag{2}$$

The upper limits of K and G are determined by the following the Voigt expressions:

$$K_{voigt} = (C_{11} + C_{22} + C_{33} + 2C_{12} + 2C_{13} + 2C_{23})/9 \tag{3}$$

$$G_{voigt} = (C_{11} + C_{22} + C_{33} + 3C_{44} + 3C_{55} + 3C_{66} - C_{12} - C_{13} - C_{23})/15 \tag{4}$$

The lower limits of K and G are determined by the following the Reuss expressions:

$$K_{Reuss} = 1/(S_{11} + S_{22} + S_{33} + 2S_{12} + 2S_{13} + 2S_{23}) \tag{5}$$

$$G_{Reuss} = 15/[4(S_{11} + S_{22} + S_{33}) - 4(S_{12} + S_{13} + S_{23}) + 3(S_{44} + S_{55} + S_{66})] \tag{6}$$

Finally, the actual values of K and G are determined by the Hill theory, averaging the corresponding values of K and G obtained from the Voigt and Reuss theories. The corresponding equations are given as follows:

$$K_{Hill} = (K_{voigt} + K_{Reuss})/2 \tag{7}$$

$$G_{Hill} = (G_{voigt} + G_{Reuss})/2 \tag{8}$$

As can be seen from Table 1, the shear moduli of the two materials, pure NR and GNS/NR, were calculated by simulation to be 1.441 GPa and 2.807 GPa, respectively. The analysis of the data results showed that when GNS was added to the natural rubber matrix as a reinforcing material, the shear modulus and bulk modulus of the composite matrix were increased by 94.8% and 110.1%, respectively. Due to the van der Waals force and the electrostatic force adsorption between GNS and the polymer matrix, the polymer chains were tightly adsorbed on the GNS surface. When the entire matrix system was subjected to the external shear force, GNS/NR required a greater shear force than the pure NR matrix to achieve the same shear strain. The addition of graphene significantly improves the shear properties and resistance to volume change of the rubber material matrix, thereby extending the service life of the rubber.

Table 1. Mechanical properties of the pure natural rubber and the graphene/natural rubber composites.

Samples	Shear Modulus/GPa	Increase (%)	Bulk Modulus/GPa	Increase (%)
Pure NR	1.441	0	1.150	0
GNS/NR	2.807	94.8	2.416	110.1

3.1.3. Simulation of Glass Transition Temperature

The glass transition temperature serves as the polymer's lower service temperature limit and reflects the material's low temperature resistance. As seen in Figure 5, there are significant discontinuities in the slope of the specific volume against temperature for the material, indicating that the rubber experiences a secondary phase transition from the glassy to the rubbery state at these discontinuities. To determine the precise glass transition temperature, a segmented linear regression was applied to all data points. The graph demonstrates that the material's specific volume increases gradually as the temperature increases, then increases quickly at the junction of the two fitted lines, showing that the rubber–glass transition occurs at the intersection point and that the temperature corresponding to the intersection point is the rubber–glass transition temperature.

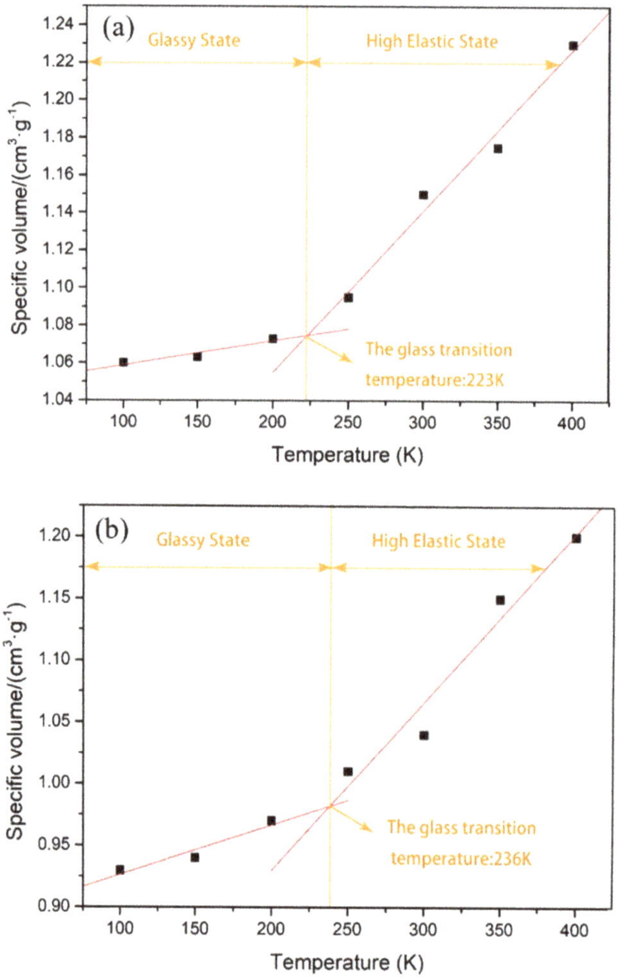

Figure 5. (**a**) Transition temperature curve for NR glass; (**b**) transition temperature curve for GNS/NR glass.

As demonstrated in Figure 5, the addition of GNS to the natural rubber matrix caused the glass transition temperature to rise by just 5.8%, from 223 K to 236 K. This suggests that the addition of GNS to the natural rubber nanocomposite matrix increases the stiffness, restricts the movement, and decreases the flexibility of the rubber molecular chain. It also increases the hardness and the glass transition temperature of the rubber, to some extent, increasing the glass transition temperature and decreasing the low temperature resistance of the rubber. However, compared to the enhancement of the rubber's mechanical qualities and tribological capabilities, the low-temperature resistance is within an acceptable range.

3.1.4. Thermal Conductivity Simulation

The non-equilibrium molecular dynamics simulation method, based on Fourier's law, was used to calculate the thermal conductivity of the two materials. The construction and optimization processes of both polymer materials were consistent with the above results. The time step was set to 1 fs, and the exchange time step was 250 at 298 K. The Berendsen thermostat was selected for temperature control, and the entire process was simulated using the COMPASS force field. The simulation results are shown in Table 2.

Table 2. Thermal conductivity of the pure natural rubber and the graphene/natural rubber composites.

Samples	Pure NR	GNS/NR
Thermal conductivity ($W{\cdot}m^{-1}{\cdot}K^{-1}$)	0.1482	0.2355

The addition of GNS increased the thermal conductivity of the natural rubber material matrix by 59%, as shown in Table 2. This is due to the fact that the addition of GNS to the interior of the natural rubber matrix increased the molecular chain orientation and the ordered structure led to an increase in the average free range of phonon diffusion and a weakening of phonon scattering, thereby enhancing the thermal conductivity of the entire rubber matrix system.

3.2. Tribological Performance Simulation

In order to study the tribological performance of the two materials, i.e., pure NR and GNS/NR, two three-layer tribological sub-models were constructed, as shown in Figure 6. Combined with the working conditions in the actual application, iron atoms were chosen as the base and top layers of the models, and a polymer model was used as the middle layer. The dimensions of both the base and top iron atom layers were 45 Å × 45 Å × 11.5 Å. The upper and lower iron atomic layers were fixed, and the two friction sub-models were optimized using the above optimization method for two amorphous molecular models. The immobilization of the upper and lower iron atomic layers was eliminated after a series of optimizations. To obtain the relative sliding tribological properties of the metallic iron atomic layer and the polymer composite configuration, a normal positive pressure load of 101 KPa was applied to the top layer, which was run for 400 ps at 0.1 Å/ps under the NVT system synthesis with a temperature of 298 K and a pressure of 101 KPa. The trajectory of the atoms in the sliding direction and the forces between the atoms were obtained and used to analyze the tribological properties.

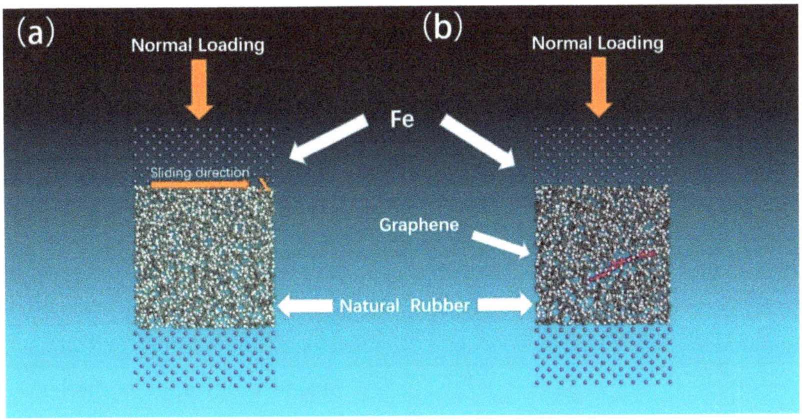

Figure 6. Molecular friction pair configuration: (**a**) pure natural rubber and iron layer model; (**b**) graphene/natural rubber and iron layer model.

To investigate the mechanism affecting the tribological properties of both the pure NR and GNS/NR nanocomposites, the profiles of atomic temperature and velocity along the thickness direction during the molecular dynamics friction simulation were calculated and extracted. The results are shown in Figures 7 and 8.

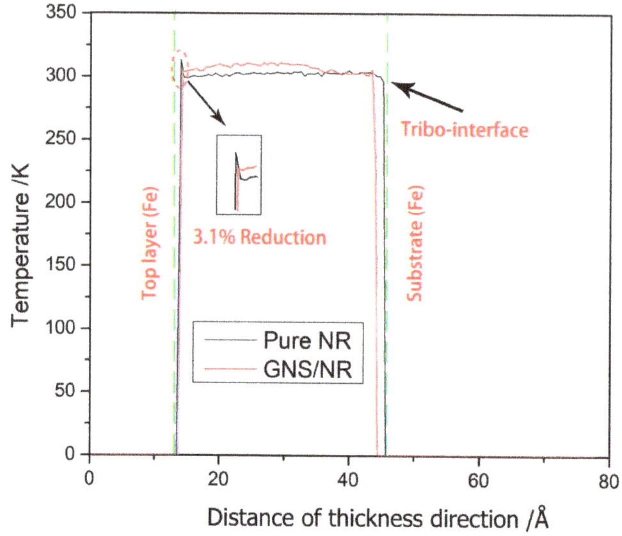

Figure 7. Temperature distribution of the pure NR and GNS/NR composites in the thickness direction.

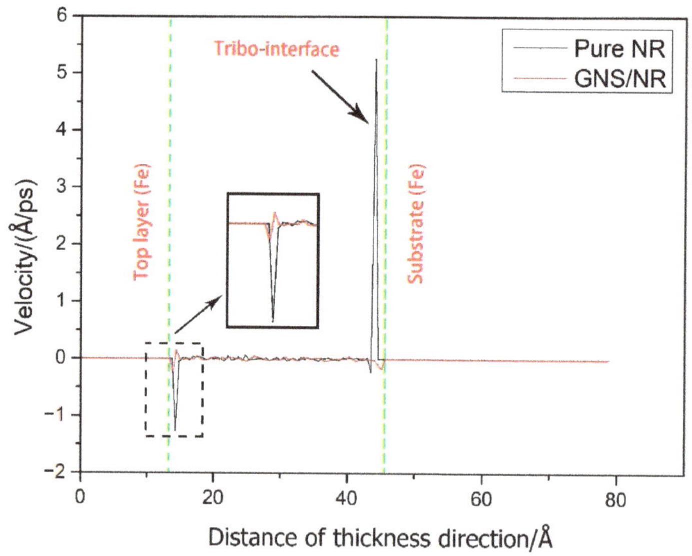

Figure 8. Velocity distribution of pure NR and GNS/NR composites in the thickness direction.

From Figure 7, for Pure NR (polyisoprene), around the top friction contact region, the temperature reached 313 K inside the material model along the thickness direction (at about 14 Å in the Z-direction); for the GNS/NR composite reinforced with GNS, the temperature was reduced by 3.1%. It has been shown that the more atoms around the friction interface between the top layer of the Fe metal and the polymer, the greater the energy dissipation. Based on Fleisher's tribological theory [24], energy dissipation occurs during the friction process, and frictional work is eventually converted into heat, leading to an increase in the temperature of the whole system. The addition of GNS reduced the interaction between the polymer molecules and the metal atoms, which lowered the coefficient of friction and eventually lowered the temperature increase generated by the entire system.

It can be further seen from Figure 8 that the pure NR substrate showed a velocity peak of 5.26 Å/ps at the friction interface, with faster molecular motion and stronger interaction with the metal Fe layer. The GNS/NR composite, on the other hand, did not show a significant velocity peak. The reason for this result is that the addition of GNS to the NR matrix increases the stability of the interface and improves the wear resistance of the composite due to the adsorption of electrostatic forces on the GNS surface and the van der Waals forces between the molecular chains and polyisoprene. As a result, the interaction with the metal Fe atomic layer was reduced, and the intermolecular motion was limited.

To obtain more insight into the wear mechanism of the two materials analyzed, Figure 9 shows snapshots of the different stages of the friction process (200 ps, 400 ps, 600 ps). From Figure 9, it is clear that pure NR, with weak mechanical properties, experiences greater deformation under shear (Figure 9a,c), fractures (Figure 9e) and finally wears away from the NR substrate. The number of atoms fractured, as shown in Figure 9e, can be identified as the wear molecule of the friction process, and the wear rate of pure NR is 6.35% under the same conditions, while the GNS/NR composite is observed to have no obvious wear phenomenon in the final state under the same conditions, and the deformation of the material can only be observed under shear (Figure 9b,d,f), further confirming that GNS has an enhancement effect on NR material wear.

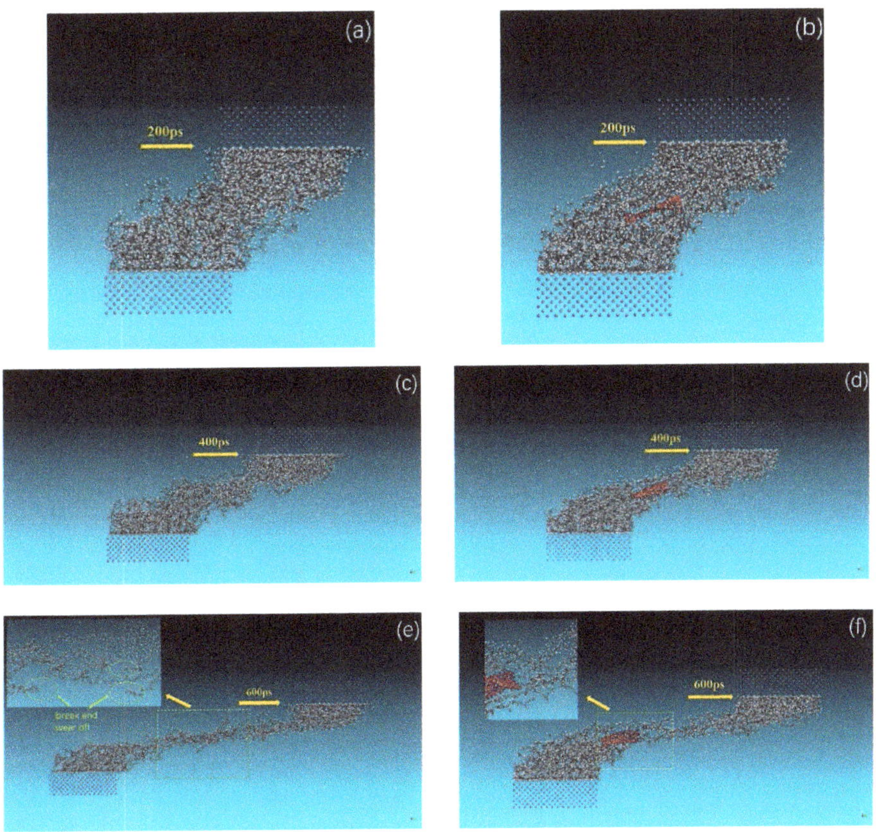

Figure 9. Friction process of the pure NR (**a**,**c**,**e**) and the GNS/NR composite (**b**,**d**,**f**) with a Fe layer at 200 ps, 400 ps, and 600 ps.

To further reveal the microscopic wear mechanism of the tribological properties of GNS-reinforced NR polymer composites, the atomic relative concentration distribution profiles along the thickness direction during the friction process were calculated for both materials, as shown in Figure 10.

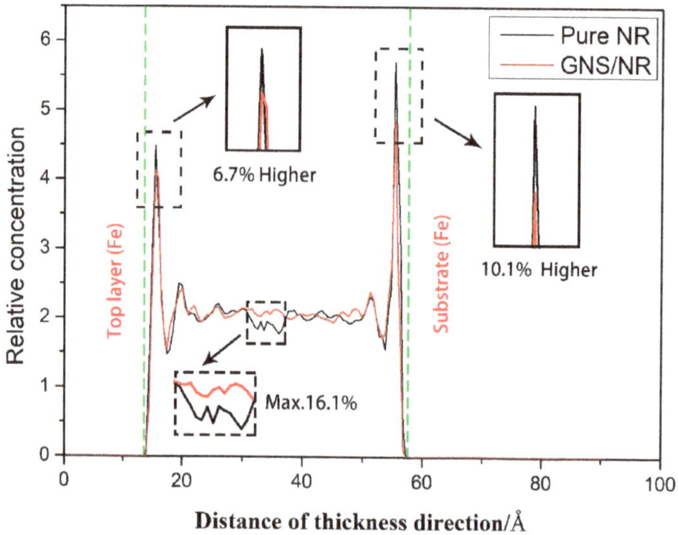

Figure 10. Relative atomic concentration distribution along the thickness direction of the pure NR and GNS/NR composites.

From Figure 10, the peaks of the relative atomic concentrations of both materials can be observed at 14 Å (top contact region) and 57 Å (bottom contact region) along the material matrix thickness direction, and the relative atomic concentrations of the pure NR composite matrix were 6.7% and 10.1% higher than those of the GNS/NR matrix near both locations, respectively. At the same time, the average relative atomic concentration of the GNS/NR matrix inside the substrate (at about 26 Å along the thickness direction) was 16.1% higher than that of the NR matrix. For the GNS/NR composite, the atoms inside the matrix were adsorbed around the GNS surface; thus, the metal Fe atomic interaction between the upper and lower layers was weakened, and the anti-friction capability was improved.

Due to VDW forces and adsorption effects [25], an interaction interface is formed between graphene and the polymer during the frictional wear process. Therefore, the interaction energy between graphene and the natural rubber matrix during the friction process is calculated and simulated based on Equation (9).

$$E_{inter} = E_{total} - E_{NR} - E_{GNS} \tag{9}$$

The frictional wear mechanism was analyzed from the perspective of energy change during material friction, and the change in the interaction potential energy between graphene and natural rubber was extracted by the frictional trajectory file; and the results are shown in Figure 11. The mutual potential energy is 17.5 kcal/mol and 145.5 kcal/mol, respectively, at the initial state (0 ps), reaching dynamic equilibrium (80 ps), and its potential energy change value increases more than 8 times, indicating the increase in interfacial interaction potential energy during frictional wear. It shows that there is a strong affinity between the GNS and NR molecular chains [26,27]. The strong interaction between the filler and the rubber leads to an enhanced adsorption between the graphene and the natural rubber matrix. The effective adsorption of the natural rubber molecular chains around the surface, which in turn leads to the reduction of natural rubber molecular chains present at the frictional interface location, reduces the interaction between the polymer matrix and the metal layer and improves the wear resistance of the natural rubber matrix.

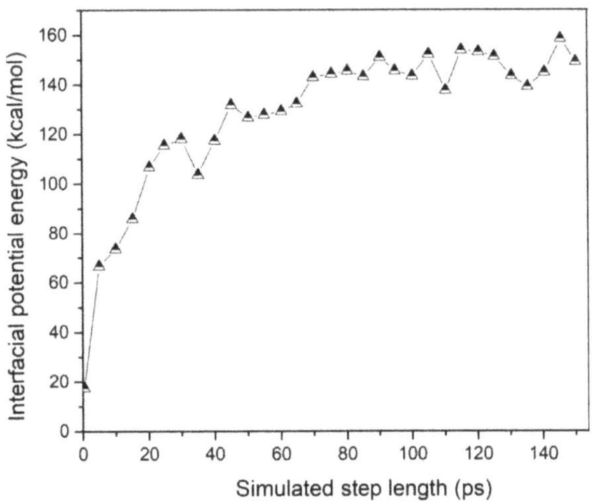

Figure 11. Graph of potential energy of GNS-NR interaction with simulation time.

4. Conclusions

Two microscopic scale molecular dynamics models pure NR and GNS/NR were developed and simulated to compare the thermodynamic and tribological properties of the two material models.

Simulations of molecular dynamics have shown that the addition of graphene significantly increases the elastic modulus, bulk modulus, and shear modulus. Significant improvements have been made to the material's tensile characteristics, resistance to deformation, and shear resistance. The enhanced orientation of the molecular chains and the ordered structure result in a weakening of the mean free range of phonon diffusion, an increase in phonon scattering, and an increase in thermal conductivity. The stiffness of the molecular chains increases, but their flexibility and resistance to low temperatures decrease.

The analysis of the microscopic information from the friction interface revealed that the atomic velocity at the friction interface between the Pure NR matrix and the metal Fe layer was faster, the atomic concentration was higher, the temperature increase at the friction interface was greater, and the system energy was higher. The addition of GNS significantly improved all the action factors at the friction interface of the composite matrix, demonstrating the enhancement of the tribological properties of the NR material matrix by GNS. This study provides a new method and theoretical basis for predicting the thermomechanical and tribological properties of inorganic nanomaterial-reinforced polymer composites.

Author Contributions: Conceptualization, X.D. and M.S.; methodology, Z.W.; software, X.Y.; validation, L.M., X.H. and J.S.; formal analysis, M.S.; investigation, M.S.; resources, L.M.; data curation, X.D.; writing—original draft preparation, X.D.; writing—review and editing, M.S.; visualization, Z.W.; supervision, L.M.; project administration, J.S.; funding acquisition, Z.W. All authors have read and agreed to the published version of the manuscript.

Funding: This research was funded by the Natural Science Foundation of Shandong Province (NO. ZR2019MEM050).

Institutional Review Board Statement: All authors have read and consented to the publication of the final draft.

Data Availability Statement: The data presented in this study are available upon request from the corresponding author.

Conflicts of Interest: The authors declare no conflict of interest.

References

1. Zhao, M.; Chen, H.; Yuan, J.; Wu, Y.; Li, S.; Liu, R. The study of ionic and entanglements self-healing behavior of zinc dimethacrylate enhanced natural rubber and natural rubber/butyl rubber composite. *J. Appl. Polym. Sci.* **2022**, *139*, 52048. [CrossRef]
2. Sethulekshmi, A.; Saritha, A.; Joseph, K. A comprehensive review on the recent advancements in natural rubber nanocomposites. *Int. J. Biol. Macromol.* **2022**, *194*, 819–842. [CrossRef] [PubMed]
3. Zhang, Z.; Zhang, Y.; Li, J.; Hassan, A.A.; Wang, S. Accelerated liquefaction of vulcanized natural rubber by thermo-oxidative degradation. *Polym. Bull.* **2022**, *79*, 1767–1786. [CrossRef]
4. Tanimura, S.; Matsuoka, T. Proton transfer in nafion membrane by quantum chemistry calculation. *J. Polym. Sci. Part B Polym. Phys.* **2004**, *42*, 1905–1914. [CrossRef]
5. Mark, J.E. *Physical Properties of Polymers Handbook*; Springer: New York, NY, USA, 2007; p. 825.
6. Ding, M.; Zou, L.; Zhang, L.; Zhao, T.; Li, Q.M. Molecular dynamics simulation of the effect of functionalized doping on the thermodynamic properties of cross-linked epoxy/carbon nanotube composites. *J. Electrotechnol.* **2021**, *36*, 5046–5057.
7. Li, Y.; Wang, S.; Arash, B.; Wang, Q. A study on tribology of nitrile-butadiene rubber composites by incorporation of carbon nanotubes: Molecular dynamics simulations. *Carbon* **2016**, *100*, 145–150. [CrossRef]
8. Li, Y.; Wang, S.; He, E.; Wang, Q. The effect of sliding velocity on the tribological properties of polymer/carbon nanotube composites. *Carbon* **2016**, *106*, 106–109. [CrossRef]
9. Li, Y.; Wang, S.; Wang, Q.; Xing, M. Molecular dynamics simulations of tribology properties of NBR (Nitrile-Butadiene Rubber)/carbon nanotube composites. *Compos. Part B* **2016**, *97*, 62–67. [CrossRef]
10. Li, Y.; Wang, Q.; Wang, S. A review on enhancement of mechanical and tribological properties of polymer composites reinforced by carbon nanotubes and graphene sheet: Molecular dynamics simulations. *Compos. Part A* **2019**, *160*, 348–361. [CrossRef]
11. Li, Y.; Wang, S.; Wang, Q.; Xing, M. Enhancement of fracture properties of polymer composites reinforced by carbon nanotubes: A molecular dynamics study. *Carbon* **2018**, *129*, 504–509. [CrossRef]
12. Hui, C.; Yan, F.; Xue, Q. Investigation of tribological properties of polyimide/carbon nanotube nanocomposites. *Mater. Sci. Eng. A* **2004**, *364*, 94–100.
13. Chen, H. Molecular Dynamics Simulation of Friction between Natural Rubber and Carbon Materials. Mater's Thesis, Qingdao University of Science and Technology, Qingdao, China, 2020.
14. Pan, K.; Huang, W.; Chen, H.; Guo, C. Molecular dynamics simulation of functionalized graphene/polydimethylsiloxane composites and their mechanical properties. *Polym. Mater. Sci. Eng.* **2020**, *36*, 97–101.
15. Song, J.; Lei, H.; Zhao, G. Improved mechanical and tribological properties of polytetrafluoroethylene reinforced by carbon nanotubes: A molecular dynamics study. *Comput. Mater. Sci.* **2019**, *168*, 131–136. [CrossRef]
16. Zhu, M.; Li, J.; Chen, J.; Song, H.; Zhang, H. Improving thermal conductivity of epoxy resin by filling boron nitride nanomaterials: A molecular dynamics investigation. *Comput. Mater. Sci.* **2019**, *164*, 108–115. [CrossRef]
17. Chawla, R.; Dhawan, M.; Sharma, S. A molecular dynamics study to predict the friction and wear behavior of carbon nanotube reinforced styrene-butadiene rubber. *Proc. Instig. Mech. Eng. Part J J. Eng. Tribol.* **2019**, *233*, 1565–1573. [CrossRef]
18. Chawla, R.; Sharma, S. A molecular dynamics study on efficient nanocomposite formation of styrene–butadiene rubber by incorporation of graphene. *Graphene Technol.* **2018**, *3*, 25–33. [CrossRef]
19. Sun, H. COMPASS: An ab Initio Force-Field Optimized for Condensed-Phase ApplicationssOverview with Details on Alkane and Benzene Compounds. *J. Phys. Chem. B* **1998**, *102*, 7338–7346. [CrossRef]
20. Rigby, D.; Sun, H.; Eichinger, B.E. Computer simulations of poly(ethylene oxide): Force field, pvt diagram and cyclization behaviour. *Polym. Int.* **1997**, *44*, 311–330. [CrossRef]
21. Li, Y.; Wang, S.; Quan, W.; Xing, M. A comparison study on mechanical properties of polymer composites reinforced by carbon nanotubes and graphene sheet. *Compos. Part B* **2017**, *133*, 35–41. [CrossRef]
22. Sun, B.; Guo, W.; Jianbing, G.U.; Liwei, M.I. Molecular Dynamics Simulation of Mechanical Properties of Graphene Oxide/Styrene Butadiene Rubber Composites. *J. Zhengzhou Univ. (Eng. Ence)* **2019**, *40*, 87–91.
23. Gojny, F.H.; Wichmann, M.H.G.; Kopke, U. Carbon nanotube-reinforced epoxy-composites: Enhanced stiffness and fracture toughness at low nanotube content. *Compos. Sci. Technol.* **2004**, *64*, 2363–2371. [CrossRef]
24. Hu, Y.; Ma, T.; Wang, H. Energy dissipation in atomic-scale friction. *Friction* **2013**, *1*, 24–40. [CrossRef]
25. Yasin, S.; Yasin, S.; Hussain, M.; Zheng, Q.; Song, Y. Influence of ionic liquid on rheological behaviors of candle soot and cellulose nanocrystal filled natural rubber nanocomposites. *Compos. Commun.* **2022**, *33*, 101214. [CrossRef]
26. Yasin, S.; Hussain, M.; Zheng, Q.; Song, Y. Large amplitude oscillatory rheology of silica and cellulose nanocrystals filled natural rubber compounds. *J. Colloid Interface Sci.* **2021**, *588*, 602–610. [CrossRef] [PubMed]
27. Yasin, S.; Hussain, M.; Zheng, Q.; Song, Y. Effects of ionic liquid on cellulosic nanofiller filled natural rubber bionanocomposites. *J. Colloid Interface Sci.* **2021**, *591*, 409–417. [CrossRef]

polymers

Article

Synthesis of Biobased Hydroxyl-Terminated Oligomers by Metathesis Degradation of Industrial Rubbers SBS and PB: Tailor-Made Unsaturated Diols and Polyols

Manuel Burelo [1,*], Selena Gutiérrez [2,*], Cecilia D. Treviño-Quintanilla [1,*], Jorge A. Cruz-Morales [2], Araceli Martínez [3] and Salvador López-Morales [4]

1 Institute of Advanced Materials for Sustainable Manufacturing, Tecnologico de Monterrey, Queretaro 76146, Mexico
2 Facultad de Química, Universidad Nacional Autónoma de México, Ciudad Universitaria, Coyoacán, Ciudad de México 04510, Mexico
3 Escuela Nacional de Estudios Superiores, Unidad Morelia, Universidad Nacional Autónoma de México, Antigua Carretera a Pátzcuaro No. 8701, Col. Ex. Hacienda de San José de la Huerta, Morelia 58190, Michoacán, Mexico
4 Instituto de Investigaciones en Materiales, Universidad Nacional Autónoma de México, Ciudad Universitaria, Coyoacán, Ciudad de México 04510, Mexico
* Correspondence: mburelo@tec.mx (M.B.); sgutierrezf@unam.mx (S.G.); cdtrevino@tec.mx (C.D.T.-Q.)

Abstract: Biobased hydroxyl-terminated polybutadiene (HTPB) was successfully synthesized in a one-pot reaction via metathesis degradation of industrial rubbers. Thus, polybutadiene (PB) and poly(styrene-butadiene-styrene) (SBS) were degraded via metathesis with high yields (>94%), using the fatty alcohol 10-undecen-1-ol as a chain transfer agent (CTA) and the second-generation Grubbs–Hoveyda catalyst. The identification of the hydroxyl groups (-OH) and the formation of biobased HTPB were verified by FT-IR and NMR. Likewise, the molecular weight and properties of the HTPB were controlled by changing the molar ratio of rubber to CTA ([C=C]/CTA) from 1:1 to 100:1, considering a constant molar ratio of the catalyst ([C=C]/Ru = 500:1). The number average molecular weight (M_n) ranged between 583 and 6580 g/mol and the decomposition temperatures between 134 and 220 °C. Moreover, the catalyst optimization study showed that at catalyst loadings as low as [C=C]/Ru = 5000:1, the theoretical molecular weight is in good agreement with the experimental molecular weight and the expected diols and polyols are formed. At higher ratios than those, the difference between theoretical and experimental molecular weight is wide, and there is no control over HTPB. Therefore, the rubber/CTA molar ratio and the amount of catalyst play an important role in PB degradation and HTPB synthesis. Biobased HTPB can be used to synthesize engineering design polymers, intermediates, fine chemicals, and in the polyurethane industry, and contribute to the development of environmentally friendly raw materials.

Keywords: poly(styrene-butadiene-styrene) (SBS); tailor-made oligomers; hydroxyl-terminated polybutadiene; unsaturated polyol; unsaturated diol; metathesis degradation; catalyst optimization

Citation: Burelo, M.; Gutiérrez, S.; Treviño-Quintanilla, C.D.; Cruz-Morales, J.A.; Martínez, A.; López-Morales, S. Synthesis of Biobased Hydroxyl-Terminated Oligomers by Metathesis Degradation of Industrial Rubbers SBS and PB: Tailor-Made Unsaturated Diols and Polyols. *Polymers* **2022**, *14*, 4973. https://doi.org/10.3390/polym14224973

Academic Editors: Hui Zhao, Wei Wu, and Bin Wang

Received: 11 October 2022
Accepted: 12 November 2022
Published: 17 November 2022

1. Introduction

Polybutadiene (PB) and poly(styrene-butadiene-styrene) (SBS) are industrial rubbers with a diverse field of applications, such as soles of shoes, sealing rings, gaskets, damping materials, insulation materials, antivibration bushes, automotive parts, and the tire industry, among others, where elastomeric properties and durability are essential [1,2]. SBS is a block copolymer whose main chain is constituted by the segments polystyrene-polybutadiene-polystyrene, and butadiene is found in a greater proportion than styrene [2,3]. Industrial PB is a homopolymer in *cis*-1,4-configuration, mainly.

The tire industry is the most important market for such synthetic rubber as SBS, SBR, PB, butyl rubber (IIR) and polyisoprene (IR). Styrene butadiene rubber is the most important grade at the moment, with around 5.1 million tons consumed in 2020. In 2020, 14.4 million tons of synthetic rubber were produced. The global demand for tires will reach 3.2 billion units in 2022. Furthermore, it is reported that a total of about 223 million replacement tires were shipped to the USA in 2019 [4,5]. It is important to note that, once their use is over or they are at the end of their life, these materials are disposed of in landfills or end up as microplastics (MPs) in the oceans. For example, the deterioration of the tires due to their abandonment or rolling produces small fragments that travel through the air, land, and water, becoming part of the MPs. Moreover, rubber-based materials are difficult to degrade naturally in landfills due to their chemical composition, high molecular weight, cross-linking, and additives, among others. Thus, large quantities of these materials accumulate in the environment, causing its deterioration and contamination [6,7].

In addressing this challenge, several waste-management methods have been investigated and tested in the last few decades, including thermal (pyrolysis), mechanical (mechanical or cryogenic trituration), physical (radiation, ultrasonic radiation, or microwave), chemical (catalyst systems, ozonolysis, free radicals, or oxidation), and biological (fungi, bacterial strains, or microbial communities) processes [6–14]. According to reports, these methods show some disadvantages. For instance, they require high temperatures and/or pressure, the use of large quantities of solvents, involve the generation of different types of toxic gases or by-products and several steps in the reactions, low yields, wide variety of products, low conversions, and most of all, there is no control over the degradation process and thus over the structure and molecular weight of the oligomers. These facts limit the use or application of the products obtained and prevent a waste-rubber recycling path.

On the contrary, the depolymerization process via metathesis has shown that both natural and industrial rubbers can be degraded under mild conditions, at low temperatures and atmospheric pressure, in solvent-free conditions, or allowing the use of essential oils as green solvents [15–19]. Metathesis allows the design of oligomers by carefully selecting the olefins used as chain transfer agents (CTA) and the polymer/CTA molar ratio. For this reason, oligomers via metathesis have been called tailor-made oligomers or telechelic oligomers due to the high control over the chemical structure, molecular weight, and functional group number per chain (Fn close to 2). Telechelic compounds, oligomers, polymers, biobased products, and raw materials with different functional groups have been obtained via metathesis degradation reaction, including hydroxyl-terminated compounds, which represent a challenging task [15–19].

Metathesis reactions are limited to compounds with double bonds [C=C] in their structure and catalysts based on transition metals such as Ru, Mo, and W. Highly active catalysts based on Ru and high concentrations of these are required to achieve the successful synthesis of hydroxy-terminated compounds [19,20]. In the last few years, the authors have reported that several factors, such as the CTA, selected catalyst, reactant concentration, solvent, and olefin ring-strain can affect the metathesis degradation reaction and synthesis of hydroxy-terminated compounds [20–22]. In this sense, unprotected alcohols used as CTAs are not fully compatible with the metathesis catalysts.

Hydroxyl-terminated polybutadiene oligomers (HTPB) can be used as diols and polyols. Diols are employed in the elaboration of adhesives, coating, sealants, intermediate products, fine chemicals, and polymeric binders for both propellants and explosives, and as chain-extender oligomers in polycondensation reactions for the synthesis of copolymers, polyether polyols, and polyurethanes (PU), where they constitute the rigid segment [23–26]. Meanwhile, polyols are used in the synthesis of polymers and copolymers by polycondensation reactions. In polyurethanes, the polyol constitutes the flexible segments in the main chain and determines the type of PU that can be obtained and its applications [27–30].

On the other hand, most of the raw materials employed for polymer production (oligomers, diols, polyols, and isocyanates) are derived from nonrenewable petroleum resources. Currently, the chemical industry and scientific community are focused mainly on replacing fossil raw materials with environmentally friendly alternatives and on developing materials that, are suitable for recycling or biodegradation at the end of the product's life [31,32]. For instance, terpenes, vegetable oils, and carbohydrates have been used as feedstock to manufacture various sustainable raw materials and products. The fatty alcohol 10-undecen-1-ol can be obtained from castor oil (*Ricinus communis*), and it has been converted into diols and polyols [33–36]. In that sense, we report the synthesis of biobased hydroxyl-terminated oligomers (HTPB) via metathesis degradation from polybutadiene and poly(styrene-butadiene-styrene) in a one-pot reaction using the fatty alcohol 10-undecen-1-ol as CTA and the second-generation Hoveyda–Grubbs catalyst Ru, as well as the catalyst optimization for this process.

2. Materials and Methods

2.1. Materials and Reagents

Cis-1,4-polybutadiene (PB) ($M_n = 9.10 \times 10^5$ g/mol, PDI = 2.20) was obtained from Mexico (Compañía Hulera, Ciudad de México, Mexico). Poly(styrene-butadiene-styrene) (SBS) (styrene, 30 wt.%) ($M_n = 1.70 \times 10^5$, PDI = 1.50), 10-undecen-1-ol (98%) (CTA), chlorobenzene anhydrous, methanol, and the catalyst [1,3-bis-(2,4,6-trimethylphenyl)-2-imidazolidinylidene]dichloro(o-isopropoxyphenylmethylene) ruthenium (second generation Hoveyda-Grubbs) (Ru) were purchased from Sigma-Aldrich, Inc. (St. Louis, MO, USA), and used as received.

2.2. Synthesis of Biobased Hydroxyl-Terminated Oligomers (Biobased HTPB) via Metathesis Degradation from SBS or PB

The synthesis of biobased HTPB was carried out via metathesis degradation reaction from industrial rubber (SBS or PB) using the fatty alcohol 10-undecen-1-ol as chain transfer agent (CTA), chlorobenzene as solvent, and second-generation Hoveyda–Grubbs catalyst Ru (Scheme 1). All reactions were performed under a nitrogen atmosphere using standard Schlenk line techniques in a 100 mL Schlenk tube containing a Spinplus stir bar from Sigma-Aldrich.

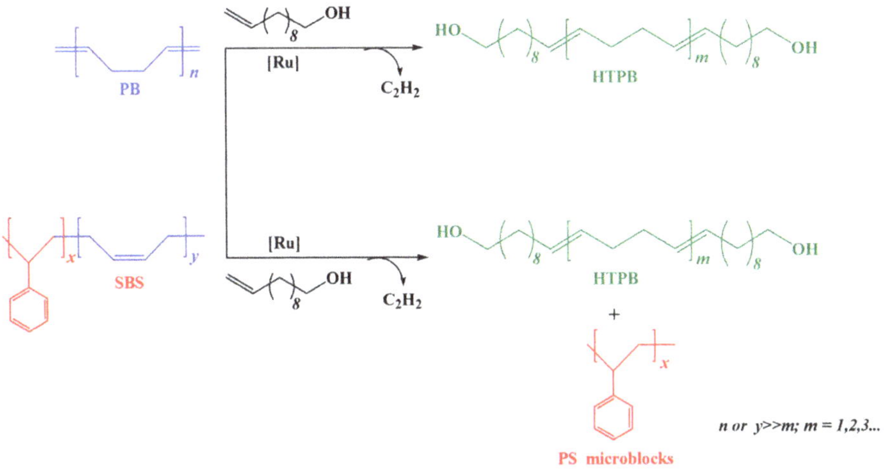

Scheme 1. Synthesis of biobased hydroxyl-terminated oligomers via metathesis degradation reaction from PB or SBS using 10-undecen-1-ol as CTA, and the Ru-alkylidene catalyst (Ru).

First, the rubber SBS or PB (3.00 g, 55.50 mmol) and CTA (9.44 g, 55.50 mmol for molar ratio rubber/CTA = 1:1) were placed in the Schlenk tube charging with chlorobenzene anhydrous as solvent (10 wt.%), then the catalyst Ru (0.13 g, 0.2220 mmol for C=C/Ru = 500) was added. The temperature was controlled using an oil bath maintained at 40 °C for 12 h with continuous stirring.

The molar ratios of carbon-carbon double bonds of SBS or PB to CTA used were [SBS]/[CTA] or [PB]/[CTA] = 1:1, 10:1, 20:1, 50:1, and 100:1 (9.44, 0.94, 0.47, 0.18 and 0.094 g, respectively) in order to control the molecular weights and for the catalyst [C=C]/[Ru] = 500. During the reaction process, ethylene is generated as a by-product, which means that is possible to displace the equilibrium process toward a specific product. Thus, in order to favor the formation of HTPB, ethylene gas was removed using a vacuum line [16,19]. After terminating the reaction by the addition of a small amount of ethyl vinyl ether (0.30 mL, 3.00 mmol), the product was precipitated three times in methanol and purified. In the case of SBS, the polystyrene microblocks (gray solid) were separated by precipitation in a methanol solution and then by decantation using a separation funnel. The product was isolated and dried under vacuum (Thermo Scientific vacuum oven, Waltham, MA, USA) at 60 °C for 8 h. Finally, the biobased HTPB was weighed and the yield calculated by the gravimetric method. The HTPBs (oligomers or polyol) were both sandy brown liquids with different molecular weights with a yield between 94% and 98%. The data are summarized in Table 1.

Table 1. Synthesis of biobased HTPB via metathesis degradation from SBS using 10-undecen-1-ol as CTA.

Entry	[SBS]/[CTA] [b] (mol/mol)	*MW* Theoretical (g/mol)	Molecular Weight [c] (g/mol)			Yield [d] (%)	Thermal Properties (TGA) (°C)	
			M_n	M_w	PDI		$T_{d\,(5\%)}$	$T_{d\,(50\%)}$
SBS [a]			170,000	255,000	1.5			
SBS1	1:1	366	583 ± 65	950 ± 80	1.63 ± 0.05	94 ± 1.00	134	400
SBS2	10:1	852	880 ± 20	1730 ± 26	1.96 ± 0.05	96 ± 1.15	147	424
SBS3	20:1	1392	1593 ± 40	3344 ± 125	2.10 ± 0.10	97 ± 0.57	152	427
SBS4	50:1	3012	3583 ± 76	9436 ± 313	2.63 ± 0.05	98 ± 0.57	158	429

[a] Molecular weights of SBS before degradation. [b] Molar ratio of double bonds [C=C] of rubber [SBS] to [CTA], using 10-undecen-1-ol as CTA. [c] Number average molecular weight (M_n), weight average molecular weight (M_w), and polydispersity index (PDI) were obtained by gel-permeation chromatography (GPC) using monodisperse polystyrene as standard. [d] Isolated yield of biobased HTPB. n = 3, means ± standard deviation.

The structure, molecular weights and thermal properties of the biobased HTPBs were characterized by infrared spectroscopy (FT-IR), nuclear magnetic resonance (NMR), gel-permeation chromatography (GPC) and thermogravimetric analysis (TGA and DTG). The reactions were carried out in triplicate and analyzed (n = 3) for each reaction using means ± standard deviation for GPG and the yields.

2.3. *Catalyst Optimization for the Synthesis of Biobased HTPB Oligomers by Metathesis Degradation*

Catalyst optimization for biobased HTPB synthesis was performed via metathesis degradation reaction of PB (3.00 g, 55.50 mmol) using the fatty alcohol 10-undecen-1-ol as CTA (9.44, 0.94 and 0.094 g) and second–generation Hoveyda–Grubbs catalyst (Ru) (Scheme 1). In all reactions, similar conditions to those described in Section 2.2 were maintained, except for the molar ratio of the catalyst, which ranged between 500–10,000. The molar ratios [C=C]/[Ru] = 500:1, 1000:1, 2000:1, 5000:1 and 10,000:1 (0.1222, 0.0611, 0.0305, 0.0122 and 0.0061 mmol of Ru, respectively) were tested in the experiments. Furthermore, the molar ratio of PB to CTA was kept constant at [PB]/[CTA] = 10:1 (0.94 g, 5.55 mmol) in the PB2, and PB4–PB7 reactions (Table 2). Thus, considering the number of repeating units of PB (m = 10), and the terminal groups from the CTA (10-undecen-1-ol), the theoretical molecular weight of oligomers is 852 g/mol (Equation (1)). After terminating the reaction, the product was precipitated three times in methanol and purified. The products were isolated and dried under a vacuum (Thermo Scientific vacuum oven, Waltham, MA, USA) at 60 °C for 8 h. Finally, the biobased HTPB was weighed and the yield calculated by the gravimetric method. The HTPBs (oligomers or polyol) were both sandy brown liquids with different molecular weights with a yield between 94% and 98%. The data are summarized in Table 2. The structure, molecular weights and thermal properties of the biobased HTPBs were characterized by NMR, GPC, and TGA. The reactions were carried out in triplicate and analyzed (n = 3) for each reaction using means ± standard deviation for GPG and the yields.

Table 2. Catalyst optimization used for the synthesis of biobased HTPB via metathesis degradation of PB.

Entry	[PB]/[CTA] [b] (mol/mol)	Molar Ratio [c] [C=C]/[Ru] (mol/mol)	Catalyst Moles (mmol)	*MW* Theoretical (g/mol)	Molecular Weight [d] (g/mol)			Yield (%) [e]	Thermal Properties (TGA, °C)	
					M_n	M_w	*PDI*		$T_{d\,(5\%)}$	$T_{d\,(50\%)}$
PB [a]					910,000	2,002,000	2.2			
PB1	1:1	500:1	0.2220	366	643 ± 21	1115 ± 58	1.73 ± 0.05	94 ± 1.00	136	402
PB2	10:1	500:1	0.1222	852	873 ± 35	1658 ± 74	1.90 ± 0.10	96 ± 1.15	150	404
PB3	100:1	500:1	0.1120	5712	6580 ± 396	14,459 ± 662	2.20 ± 0.10	98 ± 0.57	220	430
PB4	10:1	1000:1	0.0611	852	942 ±17	1916 ± 87	2.00 ± 0.05	96 ± 0.57	142	423
PB5	10:1	2000:1	0.0305	852	997 ± 15	2491 ± 95	2.50 ± 0.10	97 ± 1.15	147	424
PB6	10:1	5000:1	0.0122	852	1060 ± 40	2789 ± 64	2.60 ± 0.11	98 ± 0.57	151	426
PB7	10:1	10,000:1	0.0061	852	1753 ± 42	4385 ± 263	2.50 ± 0.10	98 ± 0.57	156	427

[a] Molecular weights of PB before degradation. [b] Molar ratio of double bonds [C=C] of rubber [PB] to [CTA], using 10-undecen-1-ol as CTA. [c] Molar ratio of catalyst, total carbon-carbon double bonds [C=C] (rubber + CTA) to catalyst [Ru]. [d] Number average molecular weight (*Mn*), weight average molecular weight (*Mw*), and polydispersity index (*PDI*) were obtained by gel-permeation chromatography (GPC) using monodisperse polystyrene as standard. [e] Isolated yield of biobased HTPB. n = 3, means ± standard deviation.

2.4. Measurements and Characterizations

Infrared spectroscopy (FT-IR) measurements of all samples were analyzed using a Perkin Elmer Frontier MIR FT-IR spectrometer (Waltham, MA, USA) fitted with a Frontier Universal Diamond/ZnSe ATR with a single reflection top plate and pressure arm. The spectra were recorded in the region of 4000–500 cm^{-1}. Data are presented as the frequency of absorption (cm^{-1}).

Nuclear magnetic resonance (^{1}H-NMR) spectra were recorded using a Bruker spectrometer at 400 MHz (Billerica, MA, USA) and deuterated solvents (chloroform-d, $CDCl_3$), and chemical shifts (δ) are reported in parts per million (ppm) relative to tetramethylsilane (TMS) as an internal standard.

The number and weight average molecular weight (M_n and M_w) and molecular weight distribution (PDI) were determined with reference to monodisperse polystyrene standards on a Waters 2695 ALLIANCE gel-permeation chromatography (GPC) instrument (Milford, MA, USA) at 30 °C with tetrahydrofuran (HPLC grade) as a solvent, a universal column, and a flow rate of 1 mL/min. At least three replicates were analyzed for each condition using means ± standard deviation for M_n, M_w and PDI.

Thermogravimetric analysis (TGA and DTG) was carried out on a TA Instrument Q5000 (New Castle, DE, USA) under a nitrogen atmosphere from room temperature to 600 °C with a heating rate of 10 °C/min.

3. Results and Discussion

3.1. Synthesis of Biobased Hydroxyl-Terminated Oligomers via Metathesis Degradation from Poly(styrene-butadiene-styrene)

In Sections 3.1–3.4 the results obtained for HTPB degrading SBS rubber and their properties are described. In Section 3.5 the results obtained for the HTPB degrading PB rubber and the catalyst optimization used in the reactions are described. The synthesis of biobased hydroxyl-terminated oligomers (biobased HTPB) was carried out via metathesis degradation reaction of SBS using the fatty alcohol 10-undecen-1-ol as CTA with different molar ratios of rubber/CTA in the presence of the Ru-alkylidene catalyst (Ru) (Scheme 1). It is worth noting that biobased HTPB can be classified as unsaturated diols and unsaturated polyols, both with the same structure but different molecular weights. According to the reactions shown in Table 1, the SBS was degraded in a controlled manner with high yields, in a range of 94–98%, considering that the SBS copolymer was constituted by microblocks of PB/PS in 70/30 wt.%.

3.2. FT-IR and NMR Analysis

FT-IR and NMR analysis confirmed the formation of HTPB. Figure 1 displays the comparative FT-IR spectra of SBS before (Figure 1a) and after their cross-metathesis degradation using the 10-undecen-1-ol as CTA (Figure 1b). In Figure 1a, the characteristic signals of SBS are observed. The absorption peak at 3025–3007 cm^{-1} was attributed to the stretching of the double bond (-CH=CH-), the absorption peaks at 2925–2851 cm^{-1} were attributed to the stretching of the aliphatic symmetric and asymmetric methylene groups ($-CH_2-$), and the absorption peaks at 1650–1436 cm^{-1} were attributed to the stretching of double bonds (C=C) present in the aliphatic chain and aromatic ring in the SBS. Finally, the signals at 1000–970 and 698 cm^{-1} are associated with the aromatic ring present in the polystyrene (PS).

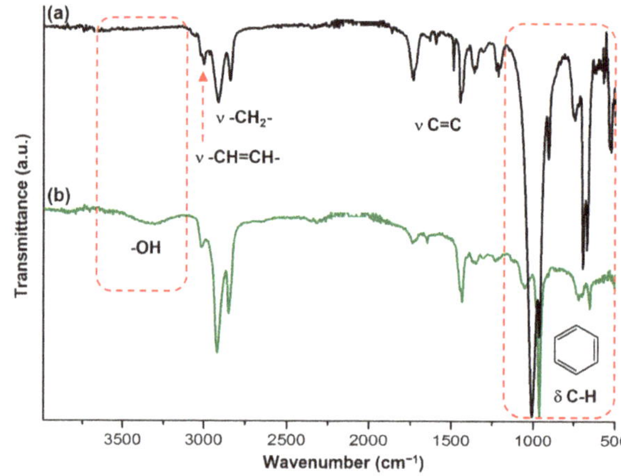

Figure 1. FT-IR spectra of (**a**) SBS and (**b**) biobased HTPB synthesized via metathesis degradation from SBS using 10-undecen-1-ol as CTA.

In contrast with Figure 1a, Figure 1b shows a broad absorption band at 3300–3250 cm^{-1}, which is associated with the R-OH stretching vibration of the HTPB. The signals of aromatic rings at 1000–970 and 698 cm^{-1} disappeared, correlated with the participation of SBS in the degradation reaction via metathesis, and the formation of the expected biobased HTPB, after isolating the microblocks of the PS. In FT-IR spectra, two features indicate the degradation in polymers: changes in signal intensity and the appearance of new signals [30,37].

The formation of biobased HTPB also was confirmed by ^1H-NMR. Figure 2 shows the comparative ^1H-NMR spectra of SBS before (Figure 2a) and after their metathesis degradation using the 10-undecen-1-ol as CTA (Figure 2b), and polystyrene isolated and recovered from the reaction (Figure 2c). In spectrum (a), the characteristic signals of SBS are observed. The signals of the aromatic ring at 7.3–6.3 ppm (from c to e), and the signals of -CH protons and methylene protons ($-CH_2-$) at 1.8–1.25 ppm (a and b) due to the aliphatic chain of styrene. The signal at 5.4 ppm (g) corresponds to the olefinic protons (-CH=CH-) of PB and the signal at 2.1–1.9 ppm (f) corresponds to methylene protons ($-CH_2-$). In spectrum (b), the signals that appeared in the range of 5.70–5.55 ppm were attributed to the formation of the double bonds (HC=CH) between the PB and 10-undecen-1-ol and the signal at 3.52 ppm (j) corresponds to the methylene protons attached to the hydroxyl groups ($-CH_2-OH$). This confirms the participation of the PB microblocks of the SBS in the metathesis degradation and the formation of biobased HTPB. In spectrum (c), the signals of the PS microblocks isolated and recovered from the reaction are observed. The signals of the aromatic ring, -CH protons, and methylene protons were previously described in the spectrum (a). As can be seen, the PS microblocks can be recovered from the metathesis degradation reaction of SBS, as shown in Scheme 1, and could be reused as oligomers or prepolymers for new polymerization reactions. NMR analysis agrees with similar spectra reported for pure PB, SBS, and natural rubber in metathesis reactions with essential oils and fatty alcohols [19,38].

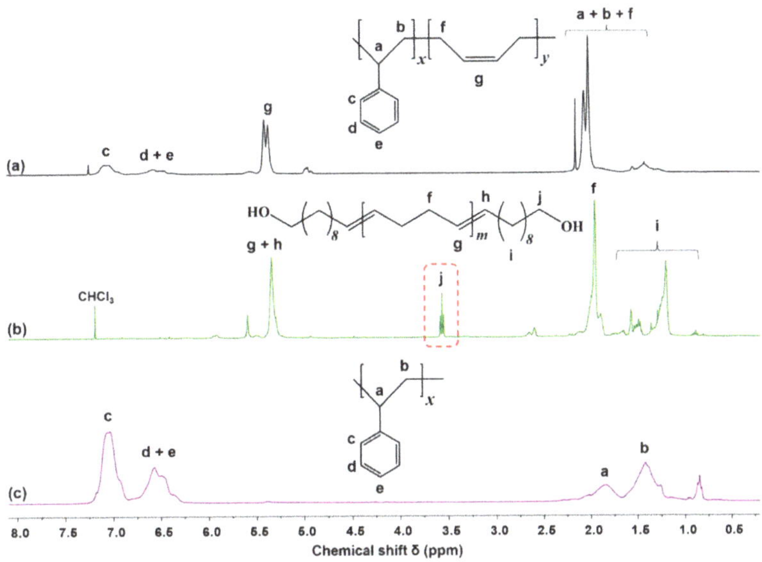

Figure 2. ^{1}H-NMR spectra (400 MHz, $CDCl_3$, TMS) of (**a**) SBS, (**b**) biobased HTPB synthesized via metathesis degradation from SBS and 10-undecen-1-ol, and (**c**) polystyrene isolated and recovered (SBS1 entry, Table 1).

3.3. Molecular Weight Analysis by GPC

The degradation process of a polymer implies that the polymeric chains are cleaved, causing a decrease in its molecular weight (M_n *or* M_w) and a change in the polydispersity index (PDI). Table 1 shows the molecular weights obtained by GPC for the biobased HTPB. The results showed that using different molar ratios of double bonds [C=C] of rubber [SBS] to [CTA] (from 1:1 to 50:1), the molecular weight of the oligomers can be controlled, considering as a reference the theoretical molecular weight. The decrease in M_n and M_w is attributed to the production of smaller chains of unsaturated diols and polyols. Thus, biobased HTPB with M_n ranging from 583 to 3583 g/mol were obtained when SBS was degraded in a controlled manner (Entries SBS1-SBS4, Table 1). It is worth mentioning that the molecular weight (M_n) of SBS before degradation was 170,000 g/mol.

Figure 3 shows the M_w and molecular weight distribution (*MWD*) of biobased HTPB. Different *MWDs* are observed, and lower molecular weights shift to the left. The decrease in M_w of SBS when this was degraded via metathesis (Entries SBS1-SBS4, Table 1) confirms the formation of oligomers. Changes and decreases in M_n, M_w and *MWD* have been reported in rubber degradation studies [19,21,39], which agrees with the changes reported in this work by GPC.

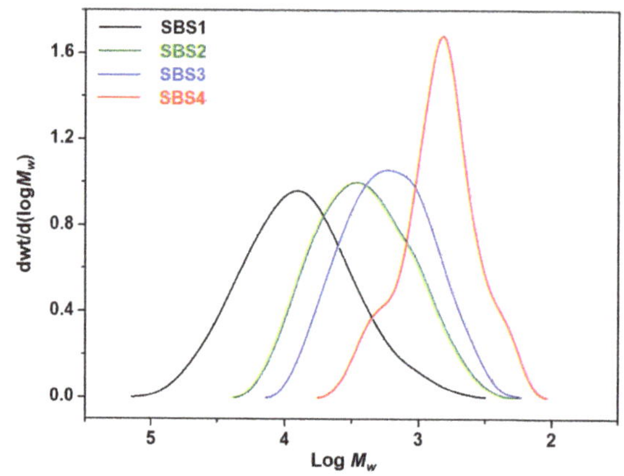

Figure 3. Molecular weight distribution (*MWD*) of biobased HTPB obtained via metathesis degradation from SBS with 10-undecen-1-ol.

3.4. Thermal Properties by TGA and DTG

Values of the decomposition temperature (T_d) of biobased HTPB formed by metathesis degradation reaction of SBS are summarized in Table 1. T_d was studied by TGA from room temperature to 600 °C, showing different percentages for mass loss (5 and 50%). Figure 4a shows TGA thermograms for the SBS oligomers, which suffer a mass loss of 5%, at 134, 147, 152, and 158 °C, for the Entries SBS1-SBS4 (Table 1), respectively. As can be seen, an increase in T_d was recorded as the molecular weight of the biobased HTPBs also increased. A similar trend was observed for T_d 50%, with temperatures ranging from 400 to 429 °C.

Figure 4b depicts the DTG thermograms from biobased HTPB obtained of SBS degradation (Entries SBS1-SBS4, Table 1), showing for SBS4 a larger area and a right shift at higher temperature values (higher molecular weight oligomer) than SBS1 (lower molecular weight oligomer). Likewise, two degradation steps were observed. The peaks observed at 150–250 °C and 400–450 °C are directly related to the decomposition of hydroxyl and olefinic bonds, respectively [33,40]. Therefore, the decomposition temperature of biobased HTPB depends on the molecular weight.

As shown earlier, the metathesis degradation reactions of polyalkenamers in the presence of unsymmetrical CTAs leads to a set of oligomers. In this sense, such olefins reduce the control over the reaction. However, this problem can be circumvented using a terminal olefin as CTA, since the metathesis is a thermodynamically controlled reaction and reaches equilibrium. As a consequence, the formation of ethylene shifts the reaction towards the products [16,19,33]. In Table 1, the 10-undecen-1-ol was used as CTA in the metathesis degradation reaction of industrial rubbers, and biobased HTPB were obtained with a high yield, thanks to the fact that ethylene formed during the metathesis degradation was removed by pressure release (Scheme 1). It has been reported that operating under reduced pressure ensures the removal of the volatile olefin [41]. Likewise, it was reported that when the 10-undecen-1-ol is used as CTA in self-metathesis or cross-metathesis reactions of PB or natural rubber, the expected products unsaturated diol and polyol were widely favored and the by-products (oligomers terminated by -OH on one end and $=CH_2$ on the other, α,ω-vinyl terminated oligomers, and ethylene) were formed with low yields. These results are in agreement with this work and other reported work for the metathesis of 10-undecen-1-ol as CTA [19,33].

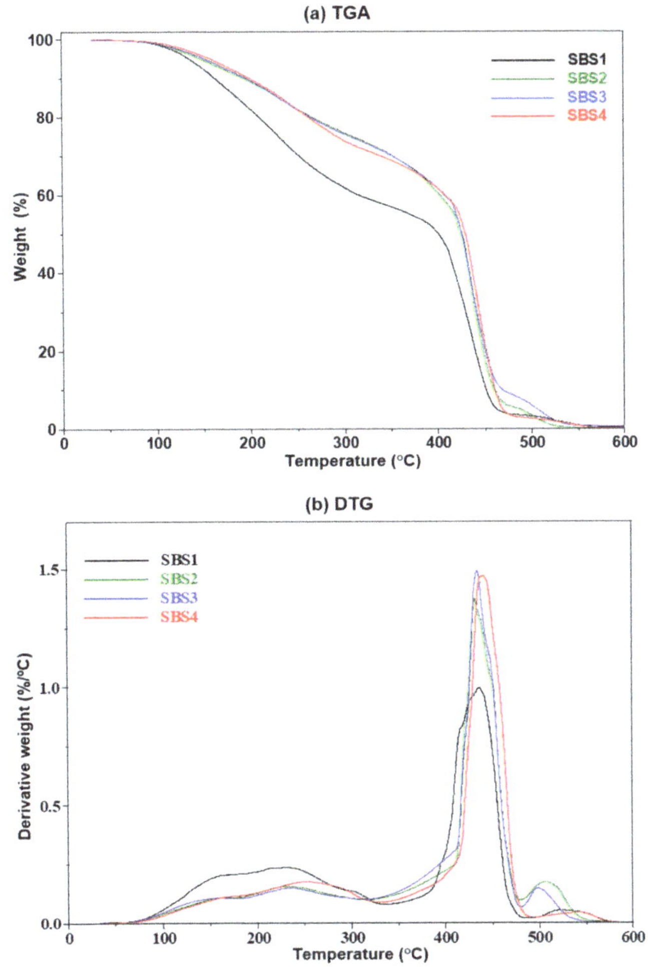

Figure 4. Thermograms of (**a**) TGA and (**b**) DTG of biobased HTPB obtained via metathesis degradation from SBS with 10-undecen-1-ol.

3.5. Catalyst Optimization for the Synthesis of Biobased HTPB via Metathesis Degradation from PB

This study synthesized biobased HTPB by metathesis degradation reaction between PB and the fatty alcohol 10-undecen-1-ol (Scheme 1). The molar ratios of the double bonds [C=C] of PB to CTA used were [PB]/[CTA] = 1:1, 10:1 and 100:1 (Table 2), keeping the molar ratio constant (10:1) in the PB2 and PB4–PB7 reactions and varying the molar ratio of Ru catalyst from [C=C]/[Ru] = 500–10,000.

According to the degradation reactions (Table 2), the PB was degraded in a controlled manner with high yields of 94–98%. It is worth mentioning that the PB industrial rubber's molecular weight (M_n) before degradation was 910,000 g/mol. Thus, biobased HTPB with M_n ranging from 643 to 6580 g/mol were obtained.

Table 2 shows that the molecular weights (M_n) obtained by GPC for the synthesized PB2 and PB4–PB7 oligomers with molar ratio constant (10:1) ranged from 873 to 1753 g/mol. The theoretical molecular weight (MW) of oligomers determined by Equation (1) was 852 g/mol.

$$Theoretical\ MW = MM_{MU} \times m + MM_{CTA} \quad (1)$$

$$MW = \frac{54g}{mol} \times 10 + \frac{312g}{mol} = 852\ g/mol$$

where,

Theoretical MW = theoretical molecular weight (g/mol).

MM_{MU} *= molecular mass of the monomeric unit, for butadiene* = 54 *g/mol*

m = repetitive units, m = 10. *It was fixed according to molar ratio PB/CTA.*

MM_{CTA} *= molecular mass of CTA* = 312 *g/mol (at both chain ends, Scheme* 1*).*

According to the results obtained using catalyst molar ratios [C=C]/[Ru] of 500–5000 (0.1222–0.0122 mmol of Ru), it was possible to control the molecular weight of oligomers (M_n). The theoretical molecular weight is very similar to the experimental one. In contrast, when a molar ratio of [C=C]/[Ru] = 10,000, the experimental molecular weight was practically twice the theoretical (Table 2). These results confirm that the molar ratio and the catalyst amount play an important role in the degradation of PB and in control of the number average molecular weight (M_n) for the synthesis of biobased hydroxyl-terminated oligomers (Figure 5).

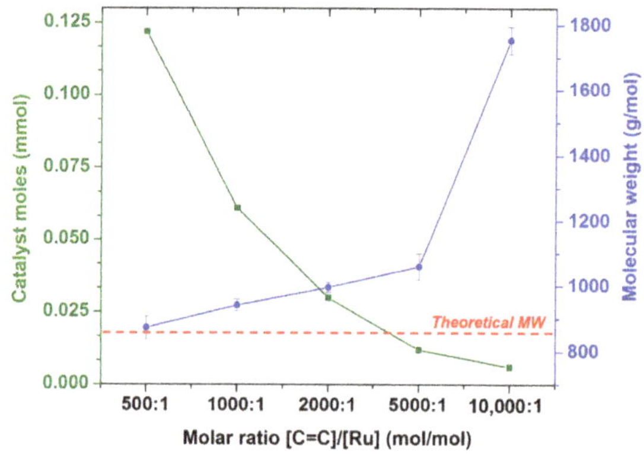

Figure 5. Relationship between molar ratio of catalyst [C=C]/[Ru] with catalyst moles and molecular weight for the synthesis of biobased HTPB. The red line indicates the theoretical molecular weight MW (852 g/mol) for all molar ratios.

NMR analysis was carried out for the biobased HTPB synthesized to verify that the degradation reactions can be controlled, i.e., the structure and molecular weight of the oligomers, changing the concentrations of the catalyst.

Figure 6 shows the spectra of PB2 and PB4–PB7 (Entries 2, 4–7, Table 2). The signals that appeared in the range of 5.7–5.40 ppm (signals g and h) were attributed to the formation of the double bonds (HC=CH) between the monomeric unit of butadiene and 10-undecen-1-ol, the signal at 2.1 ppm (f) corresponds to methylene protons (-CH_2-), and the signal at 3.52 ppm (j) corresponds to the methylene protons attached to the hydroxyl groups (-CH_2-OH). This signal (j) is essential in the spectra because—in addition to confirming the formation of oligomers—it provides a measure of control in molecular weight by functional groups. Unlike the PB7 spectrum, where a decrease in the signal (j) is observed at 3.52 ppm, this is associated with the value of repeating units of PB (*m* different from 10) and its molecular weight. Therefore, at molar ratios greater than 10,000:1, there is no control in the molecular weight of biobased HTPB.

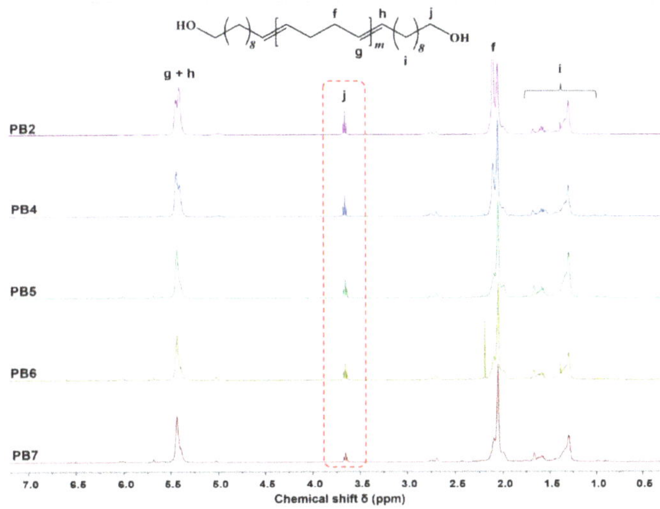

Figure 6. ^{1}H-NMR spectra (400 MHz, $CDCl_3$, TMS) of catalyst optimization used for the synthesis of biobased HTPB.

On the other hand, the T_d at 5 and 50% were obtained by TGA for the biobased HTPB obtained by metathesis degradation reaction of PB (Entries PB2, PB4-PB7, Table 2). Figure 7 shows that the temperature values and the tendency of thermograms for the biobased HTPB are similar, since they have the same structure and similar molecular weights (except PB7). Therefore, using low concentrations of Ru catalyst can be synthesized biobased HTPB with similar molecular weights and decomposition temperature (T_d).

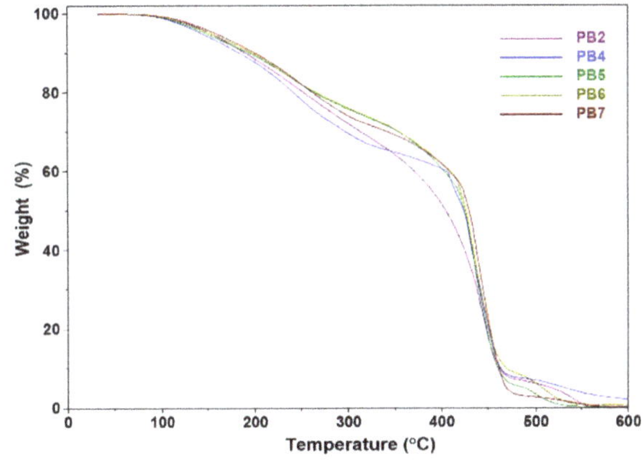

Figure 7. TGA thermograms of biobased HTPB obtained by metathesis degradation during the catalyst optimization.

Therefore, for metathesis reactions using the industrial rubbers SBS or PB, it is possible to use low concentrations of catalyst to carry out the reaction. It is reported that for metathesis reactions of rubbers, a ratio molar from 500:1–4000:1 of catalyst is used [15,19,38]. On the other hand, when natural or polyisoprene rubber is used, the reactions need an adequate concentration of catalyst, ideally from 250:1 to 1000:1 [18,19,39]. Unlike PB or SBS, natural or polyisoprene rubber requires the use of major concentrations of catalyst. This is due to their being polymers with trisubstituted unsaturations with the presence of an alkyl group directly linked to the double bond, which can impede the coordination reaction with an active metal center of the catalyst [19,22,42].

Soon, the waste derived from SBS and PB was degraded via cross-metathesis using 10-undecen-1-ol (which can be obtained from castor oil) and other natural resources, such as CTA, optimizing and economizing the catalyst. Biobased HTPB obtained with lower and higher molecular weight (tailor-made unsaturated diols and polyols, respectively) can be used to synthesize engineering design polymers, intermediates, fine chemicals, and in the polyurethane industry and contribute to developing environmentally friendly raw materials: "reuse–reduce–recycle."

4. Conclusions

Biobased HTPB (unsaturated diols or polyols by their molecular weight) were synthesized by metathesis degradation from PB and SBS using the fatty alcohol 10-undecen-1-ol as CTA and the second-generation Grubbs–Hoveyda catalyst with high yields (94–98%). The PB was degraded, as well as the PB microblocks of the SBS. Meanwhile, the polystyrene (PS) microblocks of such copolymer were isolated and recovered. Moreover, it was established that by changing the molar ratio of double bonds [C=C] of rubber (PB or SBS) to CTA in a wide range (1:1–100:1), the molecular weight and the properties could be controlled at a constant molar ratio of the catalyst ([C=C]/Ru = 500:1). The number average molecular weight (M_n) obtained ranged from 583 to 6580 g/mol, and decomposition temperatures ($T_{d\,5\%}$) from 134 to 220 °C.

The rubber/CTA molar ratio and the amount of catalyst ([C=C]/Ru) play an important role in PB degradation. The catalyst optimization study showed that at catalyst loadings as low as [C=C]/Ru = 5000:1, the theoretical molecular weight is in good agreement with the experimental molecular weight, and the expected diols and polyols are formed. The difference between theoretical and experimental molecular weights is wide at higher ratios than those, and there is no control over biobased HTPB.

Therefore, the chosen CTA and catalyst synthesized tailor-made biobased HTPBs by metathesis degradation of industrial rubbers in a one-pot reaction. The second-generation Grubbs–Hoveyda catalyst has an internal coordination bond that makes it tolerant to air, moisture, and most functional groups (-OH).

Biobased HTPB can be used to synthesize engineering design polymers, intermediates, fine chemicals and in the polyurethane industry, and contribute to the development of environmentally friendly raw materials: "reuse–reduce–recycle."

Author Contributions: Conceptualization, M.B., S.G. and C.D.T.-Q.; methodology, M.B.; investigation, M.B., C.D.T.-Q. and S.G; data curation, M.B.; formal analysis, J.A.C.-M., A.M. and S.L.-M.; writing—original draft preparation, M.B.; writing—review and editing, S.G., C.D.T.-Q., J.A.C.-M., A.M. and S.L.-M.; validation, M.B., S.G., C.D.T.-Q., J.A.C.-M., A.M. and S.L.-M.; funding acquisition, M.B., S.G., C.D.T.-Q. and A.M. All authors have read and agreed to the published version of the manuscript.

Funding: This research was funded by National Council for Science and Technology of Mexico, CONACYT (grant number 419281) and Tecnologico de Monterrey (grant number L03536294), as well as DGAPA-UNAM PAPIME (grant number PE106522).

Institutional Review Board Statement: Not applicable.

Data Availability Statement: Not applicable.

Acknowledgments: We are grateful to Gerardo Cedillo V., Karla Eriseth Reyes M., and Miguel Ángel Canseco M. (Materials Research Institute, UNAM) for their assistance with NMR, thermal analysis, and FT-IR techniques, respectively, and Adileily Olivares-Jimenez for their support in the synthesis of HTPB.

Conflicts of Interest: The authors declare no conflict of interest.

References

1. Akyüz, S.; Darı, M.E.; Esiyok, Y.E.; Ermeydan, M.A. Effects of NR/SBR Ratio on Mechanical Properties and Artificial Mechanical Performance of Anti-Vibration Bushings. *Iran. Polym. J.* **2021**, *30*, 1317–1328. [CrossRef]
2. Kanny, K.; Mohan, T.P. Rubber Nanocomposites with Nanoclay as the Filler. *Prog. Rubber Nanocompos.* **2017**, *5*, 153–177. [CrossRef]
3. Zhang, J.; Li, L.; Boonkerd, K.; Zhang, Z.; Kim, J.K. Formation of Bio-Based Elastomer from Styrene-Butadiene Copolymer and Epoxidized Soybean Oil. *J. Polym. Res.* **2014**, *21*, 404. [CrossRef]
4. Akbas, A.; Yuhana, N.Y. Recycling of Rubber Wastes as Fuel and Its Additives. *Recycling* **2021**, *6*, 78. [CrossRef]
5. Ceresana 2022© Market Report: Synthetic Rubber (E-SBR, S-SBR, BR, EPDM, IIR, NBR, CR & IR). 2022. Available online: https://www.ceresana.com/en/market-studies/plastics/synthetic-rubber/market-study-synthetic-rubber.html442 (accessed on 5 October 2022).
6. Sazali, N.; Ibrahim, H.; Jamaludin, A.S.; Mohamed, M.A.; Salleh, W.N.W.; Abidin, M.N.Z. Degradation and Stability of Polymer: A Mini Review. *IOP Conf. Ser. Mater. Sci. Eng.* **2020**, *788*, 012048. [CrossRef]
7. Nikolaivits, E.; Pantelic, B.; Azeem, M.; Taxeidis, G.; Babu, R.; Topakas, E.; Brennan Fournet, M.; Nikodinovic-Runic, J. Progressing Plastics Circularity: A Review of Mechano-Biocatalytic Approaches for Waste Plastic (Re)Valorization. *Front. Bioeng. Biotechnol.* **2021**, *9*, 696040. [CrossRef]
8. Gupte, S.L.; Madras, G. Catalytic Degradation of Polybutadiene. *Polym. Degrad. Stab.* **2004**, *86*, 529–533. [CrossRef]
9. Chakraborty, J.; Sarkar, J.; Kumar, R.; Madras, G. Ultrasonic Degradation of Polybutadiene and Isotactic Polypropylene. *Polym. Degrad. Stab.* **2004**, *85*, 555–558. [CrossRef]
10. Chen, F.; Qian, J. Studies on the Thermal Degradation of Polybutadiene. *Fuel Process. Technol.* **2000**, *67*, 53–60. [CrossRef]
11. Radhakrishnan, C.K.; Alex, R.; Unnikrishnan, G. Thermal, Ozone and Gamma Ageing of Styrene Butadiene Rubber and Poly(Ethylene-Co-Vinyl Acetate) Blends. *Polym. Degrad. Stab.* **2006**, *91*, 902–910. [CrossRef]
12. Zhang, Z.; Zhang, Y.; Li, J.; Hassan, A.A.; Wang, S. Accelerated Liquefaction of Vulcanized Natural Rubber by Thermo-Oxidative Degradation. *Polym. Bull.* **2022**, *79*, 1767–1786. [CrossRef]
13. Chittella, H.; Yoon, L.W.; Ramarad, S.; Lai, Z.W. Rubber Waste Management: A Review on Methods, Mechanism, and Prospects. *Polym. Degrad. Stab.* **2021**, *194*, 109761. [CrossRef]
14. Olejnik, T.P.; Pietras, M.; Sielski, J.; Śliżewska, K.; Sobiecka, E. The Process of Natural and Styrene–Butadiene Rubbers Biodegradation by Lactobacillus Plantarum. *Appl. Sci.* **2022**, *12*, 5148. [CrossRef]
15. Jiang, B.; Wei, T.; Zou, T.T.; Rempel, G.L.; Pan, Q.M. A Novel Approval for Degradation of Polybutadiene and Synthesis of Diene-Based Telechelic Oligomers via Olefin Cross Metathesis. *Macromol. React. Eng.* **2015**, *9*, 480–489. [CrossRef]
16. Pan, C.; Liu, P. Fluorinated Nitrile-Butadiene Rubber (F-NBR) via Metathesis Degradation: Closed System or Open System? *Eur. Polym. J.* **2022**, *162*, 110886. [CrossRef]
17. Reyes-Gómez, S.; Montiel, R.; Tlenkopatchev, M.A. Chicle Gum from Sapodilla (Manilkara zapota) as a Renewable Resource for Metathesis Transformations. *J. Mex. Chem. Soc.* **2018**, *62*, 1–15. [CrossRef]
18. Martínez, A.; Tlenkopatchev, M.A.; Gutiérrez, S.; Burelo, M.; Vargas, J.; Jiménez-Regalado, E. Synthesis of Unsaturated Esters by Cross-Metathesis of Terpenes and Natural Rubber Using Ru-Alkylidene Catalysts. *Curr. Org. Chem.* **2019**, *23*, 1356–1364. [CrossRef]
19. Burelo, M.; Martínez, A.; Cruz-Morales, J.A.; Tlenkopatchev, M.A.; Gutiérrez, S. Metathesis Reaction from Bio-Based Resources: Synthesis of Diols and Macrodiols Using Fatty Alcohols, β-Citronellol and Natural Rubber. *Polym. Degrad. Stab.* **2019**, *166*, 202–212. [CrossRef]
20. Bielawski, C.W.; Benitez, D.; Morita, T.; Grubbs, R.H. Synthesis of End-Functionalized Poly (Norbornene) s via Ring-Opening Metathesis Polymerization. *Macromolecules* **2001**, *34*, 8610–8618. [CrossRef]
21. Sedransk, K.L.; Kaminski, C.F.; Hutchings, L.R.; Moggridge, G.D. The Metathetic Degradation of Polyisoprene and Polybutadiene in Block Copolymers Using Grubbs Second Generation Catalyst. *Polym. Degrad. Stab.* **2011**, *96*, 1074–1080. [CrossRef]
22. Soares, F.A.; Steinbüchel, A. Natural Rubber Degradation Products: Fine Chemicals and Reuse of Rubber Waste. *Eur. Polym. J.* **2022**, *165*, 111001. [CrossRef]
23. Zhou, Q.; Jie, S.; Li, B.G. Preparation of Hydroxyl-Terminated Polybutadiene with High Cis-1,4 Content. *Ind. Eng. Chem. Res.* **2014**, *53*, 17884–17893. [CrossRef]
24. Chen, J.M.; Lu, Z.J.; Pan, G.Q.; Qi, Y.X.; Yi, J.J.; Bai, H.J. Synthesis of Hydroxyl-Terminated Polybutadiene Possessing High Content of 1,4-Units VIA Anionic Polymerization. *Chin. J. Polym. Sci.* **2010**, *28*, 715–720. [CrossRef]
25. Zhang, Q.; Shu, Y.; Liu, N.; Lu, X.; Shu, Y.; Wang, X.; Mo, H.; Xu, M. Hydroxyl Terminated Polybutadiene: Chemical Modification and Application of These Modifiers in Propellants and Explosives. *Cent. Eur. J. Energetic Mater.* **2019**, *16*, 153–193. [CrossRef]
26. Shahzamani, M.; Bagheri, R.; Bahramian, A.R.; Masoomi, M. Preparation and Characterization of Hybrid Aerogels from Novolac and Hydroxyl-Terminated Polybutadiene. *J. Mater. Sci.* **2016**, *51*, 7861–7873. [CrossRef]
27. Kreye, O.; Mutlu, H.; Meier, M.A.R. Sustainable Routes to Polyurethane Precursors. *Green Chem.* **2013**, *15*, 1431–1455. [CrossRef]
28. Chen, G.; Liang, Y.; Xiang, D.; Wen, S.; Liu, L. Relationship between Microstructure and Dielectric Property of Hydroxyl-Terminated Butadiene–Acrylonitrile Copolymer-Based Polyurethanes. *J. Mater. Sci.* **2017**, *52*, 10321–10330. [CrossRef]
29. Sundang, M.; Nurdin, N.S.; Saalah, S.; Singam, Y.J.; Al Edrus, S.S.O.; Ismail, N.M.; Sipaut, C.S.; Abdullah, L.C. Synthesis of Jatropha-Oil-Based Polyester Polyol as Sustainable Biobased Material for Waterborne Polyurethane Dispersion. *Polymers* **2022**, *14*, 3715. [CrossRef]

30. Burelo, M.; Gaytán, I.; Loza-Tavera, H.; Cruz-Morales, J.A.; Zárate-Saldaña, D.; Cruz-Gómez, M.J.; Gutiérrez, S. Synthesis, Characterization and Biodegradation Studies of Polyurethanes: Effect of Unsaturation on Biodegradability. *Chemosphere* **2022**, *307*, 136136. [CrossRef]
31. Hoong, S.S.; Yeong, S.K.; Hassan, H.A.; Din, A.K.; Choo, Y.M. Synthesis and Characterization of Polyurethanes Made from Copolymers of Epoxidized Natural Oil and Tetrahydrofuran. *J. Oleo Sci.* **2015**, *64*, 101–115. [CrossRef]
32. Peyrton, J.; Chambaretaud, C.; Sarbu, A.; Avérous, L. Biobased Polyurethane Foams Based on New Polyol Architectures from Microalgae Oil. *ACS Sustain. Chem. Eng.* **2020**, *8*, 12187–12196. [CrossRef]
33. Fonseca, L.R.; Bergman, J.A.; Kessler, M.R.; Madbouly, S.A.; Lima-Neto, B.S. Self-Metathesis of 10-Undecen-1-Ol with Ru-Amine-Based Complex for Preparing the Soft Segment and Chain Extender of Novel Castor Oil-Based Polyurethanes. *Macromol. Symp.* **2016**, *368*, 30–39. [CrossRef]
34. Sardon, H.; Mecerreyes, D.; Basterretxea, A.; Avérous, L.; Jehanno, C. From Lab to Market: Current Strategies for the Production of Biobased Polyols. *ACS Sustain. Chem. Eng.* **2021**, *9*, 10664–10677. [CrossRef]
35. Firdaus, M.; Montero De Espinosa, L.; Meier, M.A.R. Terpene-Based Renewable Monomers and Polymers via Thiol-Ene Additions. *Macromolecules* **2011**, *44*, 7253–7262. [CrossRef]
36. Saha, P.; Khomlaem, C.; Aloui, H.; Kim, B.S. Biodegradable Polyurethanes Based on Castor Oil and Poly (3-Hydroxybutyrate). *Polymers* **2021**, *13*, 1387. [CrossRef] [PubMed]
37. Gaytán, I.; Sánchez-Reyes, A.; Burelo, M.; Vargas-Suárez, M.; Liachko, I.; Press, M.; Sullivan, S.; Cruz-Gómez, M.J.; Loza-Tavera, H. Degradation of Recalcitrant Polyurethane and Xenobiotic Additives by a Selected Landfill Microbial Community and Its Biodegradative Potential Revealed by Proximity Ligation-Based Metagenomic Analysis. *Front. Microbiol.* **2020**, *10*, 2986. [CrossRef] [PubMed]
38. Martínez, A.; Zuniga-Villarreal, N.; Gutierrez, S.; Tlenkopatchev, M.A. New Ru-Vinylidene Catalysts in the Cross-Metathesis of Natural Rubber and Poly(Styrene-Co-Butadiene) with Essential Oils. *Curr. Org. Synth.* **2016**, *13*, 876–882. [CrossRef]
39. Hu, G.; Lin, S.; Zhao, B.; Pan, Q. Synthesis and Characterization of Natural Rubber-Based Telechelic Oligomers via Olefin Metathesis. *J. Appl. Polym. Sci.* **2020**, *49899*, e49899. [CrossRef]
40. De Vasconcelos Vieira Lopes, R.; Loureiro, N.P.D.; Pezzin, A.P.T.; Gomes, A.C.M.; Resck, I.S.; Sales, M.J.A. Synthesis of Polyols and Polyurethanes from Vegetable Oils-Kinetic and Characterization. *J. Polym. Res.* **2013**, *20*, 238. [CrossRef]
41. Astruc, D. The Metathesis Reactions: From a Historical Perspective to Recent Developments. *New J. Chem.* **2005**, *29*, 42–56. [CrossRef]
42. Fomine, S.; Tlenkopatchev, M.A. Computational Modeling of Renewable Molecules. Ruthenium Alkylidene-Mediated Metathesis of Trialkyl-Substituted Olefins. *Organometallics* **2010**, *29*, 1580–1587. [CrossRef]

polymers

MDPI

Article

Nanohybrid of Co_3O_4 Nanoparticles and Polyphosphazene-Decorated Ultra-Thin Boron Nitride Nanosheets for Simultaneous Enhancement in Fire Safety and Smoke Suppression of Thermoplastic Polyurethane

Yizhang Tong [1,†], Wei Wu [2,†], Wanjing Zhao [1], Yurui Xing [3], Hongti Zhang [3], Cheng Wang [4], Timothy B. Y. Chen [4], Anthony C. Y. Yuen [4], Bin Yu [5], Xianwu Cao [1,*] and Xiaohong Yi [2,*]

1 Key Laboratory of Polymer Processing Engineering of Ministry of Education, Guangdong Provincial Key Laboratory of Technique and Equipment for Macromolecular Advanced Manufacturing, School of Mechanical and Automotive Engineering, South China University of Technology, Guangzhou 510640, China
2 Jihua Laboratory, Foshan 528200, China
3 School of Physical Science and Technology, Shanghai Tech University, Shanghai 201210, China
4 School of Mechanical and Manufacturing Engineering, University of New South Wales, Sydney 2052, Australia
5 State Key Laboratory of Fire Science, University of Science and Technology of China, Hefei 230026, China
* Correspondence: ppeme@scut.edu.cn (X.C.); yixh@jihualab.ac.cn (X.Y.)
† The authors contribute equally to this work.

Citation: Tong, Y.; Wu, W.; Zhao, W.; Xing, Y.; Zhang, H.; Wang, C.; Chen, T.B.Y.; Yuen, A.C.Y.; Yu, B.; Cao, X.; et al. Nanohybrid of Co_3O_4 Nanoparticles and Polyphosphazene-Decorated Ultra-Thin Boron Nitride Nanosheets for Simultaneous Enhancement in Fire Safety and Smoke Suppression of Thermoplastic Polyurethane. *Polymers* **2022**, *14*, 4341. https://doi.org/10.3390/polym14204341

Academic Editor: Hui Zhao

Received: 27 September 2022
Accepted: 11 October 2022
Published: 15 October 2022

Abstract: Thermoplastic polyurethane (TPU) is widely used in daily life due to its characteristics of light weight, high impact strength, and compression resistance. However, TPU products are extremely flammable and will generate toxic fumes under fire attack, threatening human life and safety. In this article, a nanohybrid flame retardant was designed for the fire safety of TPU. Herein, Co_3O_4 was anchored on the surface of exfoliated ultra-thin boron nitride nanosheets (BNNO@Co_3O_4) via coprecipitation and subsequent calcination. Then, a polyphosphazene (PPZ) layer was coated onto BNNO@Co_3O_4 by high temperature polymerization to generate a nanohybrid flame retardant named BNNO@Co_3O_4@PPZ. The cone calorimeter results exhibited that the heat release and smoke production during TPU combustion were remarkably restrained after the incorporation of the nanohybrid flame retardant. Compared with pure TPU, the peak heat release rate (PHRR) decreased by 44.1%, the peak smoke production rate (PSPR) decreased by 51.2%, and the peak CO production rate (PCOPR) decreased by 72.5%. Based on the analysis of carbon residues after combustion, the significant improvement in fire resistance of TPU by BNNO@Co3O4@PPZ was attributed to the combination of quenching effect, catalytic carbonization effect, and barrier effect. In addition, the intrinsic mechanical properties of TPU were well maintained due to the existence of the PPZ organic layer.

Keywords: thermoplastic polyurethane; boron nitride; Co_3O_4; polyphosphazene; fire safety; mechanical property

1. Introduction

Thermoplastic polyurethane (TPU) elastomer is a type of polymer possessing superior toughness and excellent shock and corrosion resistance [1–5]. It has been utilized in wide-ranging applications in many areas, such as automotives, electronic appliances, daily necessities, etc. However, TPU is highly flammable and it is usually accompanied by a large amount of smoke and toxic gas release during its combustion, which limits the further applications of TPU and causes harm to the environment and human body [6–10]. Various types of functionalized 2D nanofillers have been incorporated into the TPU matrix to enhance its fire resistance. Yu et al. [11] utilized cetyltrimethyl ammonium bromide and tetrabutyl phosphine chloride to modify $Ti_3C_2T_x$ ultra-thin nanosheets so as to achieve the remarkable dispersion of $Ti_3C_2T_x$ in TPU. The functionalized $Ti_3C_2T_x$ was able to greatly reduce the smoke product rate (SPR) and heat release rate (HRR) by more than 50%. Huang et al. [12] introduced phosphorus-containing HBPSi onto graphene oxide (GO) and mixed this novel type of flame retardant with TPU, which dramatically inhibited peak heat release rate (PHRR) and peak smoke product rate (PSPR) by 63.5 and 58.3% during combustion. The joint effects of the phosphorus radical scavenger, the barrier effect of GO, and the three-dimensional Si-O-Si framework of P-HBPSi played important roles in the improvement of the fire safety of TPU. Furthermore, MoS_2 [13] and layered double hydroxide (LDH) were also utilized as effective flame retardants for TPU nanocomposites [14–16].

Recently, there has been a continuous interest in hexagonal boron nitride (h-BN) because of its unique features, such as high thermal stability, excellent chemical resistance, and superior thermal conductivity [17,18]. Thanks to its ultra-high temperature stability and unique 2D layered structure, BN can act as a barrier to inhibit the rapid heat and mass transfer of the polymer and delay the release of pyrolysis gas in the combustion of polymers [19–21]. Up to now, h-BN has been explored as a novel type of flame retardant filler for TPU [22–27]. Cai et al. [24] constructed a type of novel hybrid flame retardant for TPU based on SiO_2, phytic acid (PA), and h-BN (h-BN@SiO_2@PA). Compared with h-BN, the hybrid flame retardant h-BN@SiO_2@PA overcame the chemical inertia of h-BN, achieving a better flame retardant effect and toxic gas suppression while greatly enhancing the mechanical properties of the TPU nanocomposite. Generally, the flame retardant efficiency of single flame retardants is not high, while the compounding of multiple types of flame retardants or constructing hybrid flame retardants can achieve high flame retardant efficiency and reduce the loading of the flame retardant as well as maintaining other properties of TPU [28–31]. However, the lip–lip interaction [32] between the B atom and N atom in adjacent layers of h-BN has a poor interfacial interaction with polymer matrices, which seriously affects the flame retardant efficiency of BN and deteriorates the mechanical properties of polymer nanocomposites. To overcome these shortcomings, it is necessary to functionalize the surface of the h-BN [33–37].

Polyphosphazene (PPZ) is a kind of polymer with an organic–inorganic hybrid structure, which can be prepared by the thermal ring opening polymerization or solution ring opening polymerization of Hexachlorocyclotriphosphazene (HCCP) [38]. Due to the existence of P and N elements and excellent thermal stability, PPZ has been utilized to improve the fire resistance of polymers [39–42]. Singh et al. [39] directly blended PPZ with TPU and its flame retardancy was evaluated by a limiting oxygen index (LOI) analyzer and UL-94 (vertical burning test), showing the rating of V-0 with an achieved LOI value of 31.4% for the TPU/PPZ blend. Qiu et al. [41] deposited PPZ on the surface of MoS_2 nanosheets using thermal ring opening polymerization, which significantly reduced the PHRR and total heat release (THR) of EP. Furthermore, it has been proven that the combination of PPZ and transition metal elements can more effectively improve the fire resistance of flame retardants [43–46].

In this work, the exfoliated boron nitride nanosheets (BNNO) were decorated with Co_3O_4 nanoparticles and a PPZ layer to obtain a novel nanohybrid flame retardant. It was expected that the barrier effect of BNNO, combined with the promoting dispersion effect, the catalytic charring effect, and combustible gas dilution effect of Co_3O_4 and PPZ, was able to improve the flame retardant efficiency of the TPU composite. The effects of the flame retardant fillers on the morphology and mechanical performance of TPU nanocomposites were also investigated. Furthermore, the flame retardant mechanism of TPU/BNNO@Co_3O_4@PPZ nanocomposite was illustrated in detail.

2. Materials and Methods

2.1. Materials

Thermoplastic polyurethane (TPU, WHT-1570IC) with a density of 1.15 g/cm^3 was purchased from Wanhua Chemical Group Co., Ltd. (Yantai, China). The hexagonal boron nitride (h-BN, purity > 99.5%) was obtained from Qingzhou Materials Co., Ltd. (Qingzhou, China). Cobalt (II) acetate tetrahydrate ($C_4H_6CoO_4 \cdot 4H_2O$, 99.5%) was obtained from Shanghai Macklin Biochemical Co. Ltd., Shanghai, China. HCCP (purity, 98%) were supplied by J&K Chemical Co., Ltd. (Shanghai, China). Isopropanol (purity > 99.7%) and ammonia ($NH_3 \cdot H_2O$, 25 wt%) were supplied by Guangzhou Chemical Reagent Factory.

2.2. Preparation of BNNO@Co_3O_4@PPZ

The preparation routes of BNNO, BNNO@Co_3O_4, and BNNO@Co_3O_4@PPZ are illustrated in Figure 1. To achieve a well dispersion in isopropanol, h-BN was first treated at a high temperature to obtain hydroxyl groups on the h-BN surface. A certain amount of h-BN was put in a quartz boat and inserted into the center of a GSL-1100X-S tubular furnace (Hefei Kejing Materials Technology Co. Ltd., Hefei, China). Then the system temperature was raised to 1000 °C at 20 °C/min, followed by an isothermal process for 2 h in air atmosphere. The as-prepared oxidized BN was dispersed in 10 mL of isopropanol and then exfoliated with tip ultrasonication (25 kHz, 900 W) for 45 min. The resulting sample was denoted as BNNO.

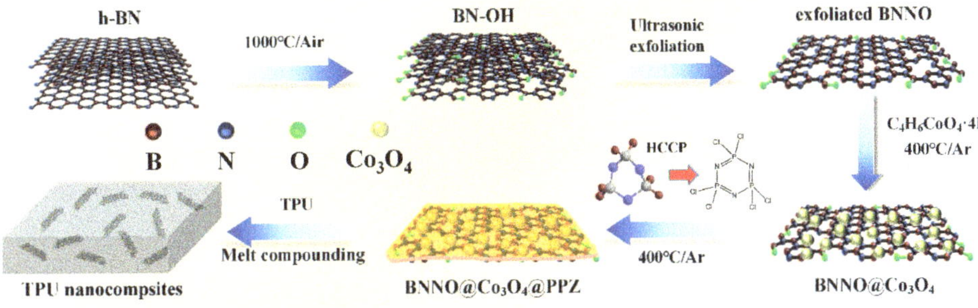

Figure 1. Preparation route of TPU/BNNO@Co_3O_4@PPZ nanocomposites.

The preparation method of Co_3O_4-decorated BNNO (BNNO@Co_3O_4) was based on previous studies [47]. First, 150 mg of BNNO was dispersed into 200 mL of DI water under tip sonication for 1 h. Then, 200 mg of $C_4H_6CoO_4 \cdot 4H_2O$ was mixed with 200 mL of DI water and transformed into the three-necked flask together with the dispersed BNNO dispersion. An amount of 7 mL of ammonia solution was added dropwise into the suspension. Then, the mixture was continuously stirred under condensation reflux for 4 h at 100 °C to ensure complete ion adsorption. Finally, the target BNNO@Co_3O_4 was obtained by thermal treatment at 400 °C in a muffle furnace.

The in situ synthesis of the polyphosphazene layer coated BNNO@Co_3O_4 was prepared via high temperature polymerization [41]. Typically, the BNNO@Co_3O_4 (2.5 g) was dispersed in isopropanol under ultrasonication for 1 h to form a homogeneous suspension. The 0.5 g of HCCP was added into the above solution by sonication. After that, the solution was stirred continuously at 70 °C to continue reacting for 2 h, and the mixture was dried at 90 °C to remove the isopropanol. The obtained solid was annealed in a tubular furnace at 700 °C for 3 h to synthesize a polyphosphazene layer. The resulting powder was centrifuged with isopropanol three times and abbreviated as BNNO@Co_3O_4@PPZ.

2.3. Preparation of TPU Nanocomposites

The TPU nanocomposites containing 2 wt% BNNO, BNNO@Co_3O_4, and BNNO@Co_3O_4 @PPZ, respectively, were prepared by mixing TPU pellets with filler powder using a Brabendermixer at 170 °C for 10 min. Then the mixtures were made into different size specimens by hot-compression at 190 °C for various testing.

2.4. Characterization

The morphologies of the BNNO, BNNO@Co_3O_4, BNNO@Co_3O_4@PPZ, fracture surfaces of TPU nanocomposites, and micro-morphology of char residues were analyzed by a scanning electron microscope (SEM, FEI Quanta 250). The morphology and element distribution of fillers were measured on a transmission electron microscopy (TEM, JEOL JEM-2100F) equipped with an energy-dispersive X-ray spectrometer (EDX). The thickness of exfoliated BNNO was determined by atomic force microscopy (AFM, Veeco Multimode V) in the tapping mode. Fourier transform infrared spectroscopy (FT-IR) spectra were conducted on a FT-IR spectrometer (Nicolet iS50) in the wavenumber range from 400 to 2500 cm^{-1}. X-ray diffraction (XRD) patterns were obtained from an X-ray diffractometer (AXS D2, Bruker, Germany) with Cu-Ka radiation. The elemental composition of fillers and char residue were analyzed by an X-ray photoelectron spectrometer (XPS, Kratos Axis Supra+) with a monochromatic Al Kα X-ray source (1486.6 eV) used for analysis. The thermal behavior of TPU nanocomposites was measured by a thermogravimetric analyzer (TGA, Netzsch TG 209 F1). The specimens of about 10 mg were heated from room temperature to 800 °C at a ramping rate of 10 °C/min under N_2 atmosphere. The mechanical properties of TPU nanocomposites were measured by an electronic universal testing machine (Instron Model 5566). The dumbbell-shaped specimens were tested at a fixed speed of 200 mm/min. The reported values were the averages of five specimens. The real fire performance of TPU nanocomposites were measured on a cone calorimeter (i-CONE, Fire Testing Technology). The dimensions of the squared specimens was 100 × 100 × 3 mm^3. The samples wrapped with aluminum foil were heated at a 35 kW/m^2 external heat flux. The structure of the char residues after cone tests were tested on a RAMANLOG 6 laser Raman spectrometer equipped with a 532 nm laser. Thermogravimetric analyzer-Infrared spectroscopy (TG-IR) was executed with a METTLER TOLEDO TGA2 thermogravimetric analyzer coupled with a Thermofisher iS50 FTIR spectrophotometer under air atmosphere from 30 to 700 °C at a ramping rate of 20 °C/min.

3. Results

3.1. Characterization of BNNO@Co_3O_4@PPZ Hybrids

The microscopic morphology of exfoliated BNNO, BNNO@Co_3O_4, and BNNO@Co_3O_4 @PPZ is shown in Figure 2. The AFM image in Figure 2a demonstrates that the BNNO with a 2–3 nm thickness is few-layer, indicating that the exfoliation of h-BN is successful [48]. In Figure 2b, the exfoliated BNNO exhibits a typical smooth lamellar structure with dimensions of around 5 μm. It should be noted that Co_3O_4 nanoparticles are found on the surface of BNNO@Co_3O_4, as shown in Figure 2c. After high-temperature polymerization, the PPZ layer has been deconjugated on the BNNO and covers the Co_3O_4 nanoparticles, which can be clearly seen in Figure 2d. The TEM image and EDX element mapping images in Figure S1 demonstrate that the existence of P and Co elements is uniformly distributed on the surface of BNNO@Co_3O_4@PPZ.

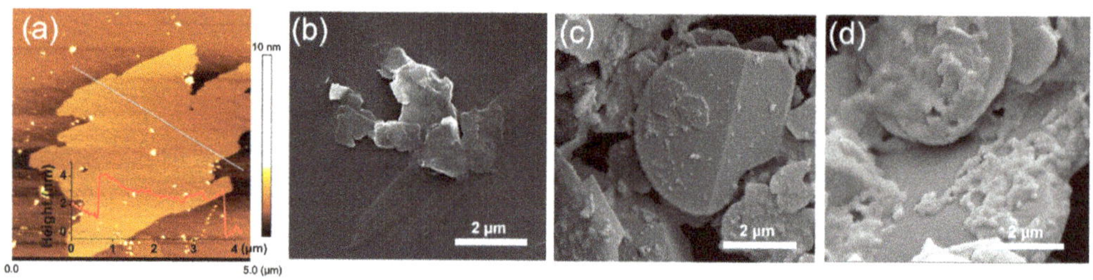

Figure 2. (**a**) AFM image (insert is thickness curve of the white line), SEM images of (**b**) BNNO, (**c**) BNNO@Co_3O_4, (**d**) BNNO@Co_3O_4@PPZ.

The FT-IR spectra of exfoliated BNNO, BNNO@Co_3O_4, and BNNO@Co_3O_4@PPZ are shown in Figure 3a. In the spectrum of BNNO, the characteristic peaks bands of boron nitride at 1386 and 806 cm^{-1} can be observed, which are ascribed to the vibrations of the B-N bond. In the spectrum of BNNO@Co_3O_4, the absorption bands belong to the bridging vibration of O-Co-O and the stretching vibration of Co-O, which appear at 663 and 569 cm^{-1}, respectively [49,50]. After the polymerization of PPZ, the strong bands at 940 and 1115 cm^{-1} are attributed to the stretching vibration of P-N and P = N bands in cyclotriphosphazene, indicating the existence of PPZ on the BNNO surface [41]. The XRD pattern of BNNO in Figure 3b shows multiple diffraction peaks at 26.8°, 41.8°, 43.5°, 50.2°, and 55.1°, corresponding to the BN crystal planes (002), (100), (101), (102), and (004), respectively [22]. After the decoration of Co_3O_4, several typical diffraction peaks at 19.0°, 31.2°, 36.5°, 44.8°, and 59.3° can be seen, which correspond to the (111), (220), (311), (400), and (511) planes of Co_3O_4. These characteristic peaks are well matched to the standard diffraction pattern of Co_3O_4 (JCPDS: 43-1003) [49]. This indicates that the Co_3O_4 is successfully decorated on the BNNO surface. In addition, there are some small peaks in the pattern of BNNO@Co_3O_4@PPZ, which originate from the amorphous PPZ. The intensity of the characteristic peaks of Co_3O_4 is reduced after coating the PPZ layer, which is consistent with other previous works [51,52]. Figure 3c shows the XPS full survey spectra of BNNO, BNNO@Co_3O_4, and BNNO@Co_3O_4@PPZ. Several intense peaks at 59, 99, and around 775–805 eV can be observed in the XPS spectrum of BNNO@Co_3O_4, which are ascribed to the Co 3p, Co 3s, and Co 2p peaks of Co_3O_4, indicating the occurrence of Co_3O_4 on the surface of BNNO [49]. To gain more insight, the XPS spectrum from 775 to 805 eV is enlarged in Figure S2, and the Co 2p exhibits two split peaks of Co $2p_{3/2}$ and Co $2p_{1/2}$, which appear at around 779 and 795 eV, respectively. A new peak of P 2p is clearly observed in the spectrum of BNNO@Co_3O_4@PPZ, confirming the successful coating of the PPZ layer on BNNO@Co_3O_4. In the high-resolution spectrum of P 2p peak in Figure 3d, the peak of P 2p can be fitted into three peaks, which are located at 132.2 eV (P $2p_{3/2}$ of P species), 133 eV (P = N bond), and 133.7 eV (P-O), respectively. As can be seen in Figure S3, BNNO, BNNO@Co_3O_4, and BNNO@Co_3O_4@PPZ have no obvious mass loss until 800 °C. Based on the above analysis, it can be reasonably proven that the BNNO@Co_3O_4@PPZ has been prepared successfully.

3.2. Fracture Surface Morphology of TPU Nanocomposites

Figure 4 shows the SEM graphs of the brittle fracture surface of pure TPU, TPU/BNNO, TPU/BNNO@Co_3O_4, and TPU/BNNO@Co_3O_4@PPZ. In Figure 4a,b, pure TPU is brittle damaged with a smooth fracture surface. With the addition of BNNO and BNNO@Co_3O_4, the fracture surfaces displays an uneven condition, as shown in Figure 4c,e. Although there is no obvious agglomeration in the field of vision, as shown in Figure 4d,f, the debonding phenomenon occurs when observing the interface of the filler–matrix, indicating relatively poor compatibility [53]. In contrast, few fillers are exposed to the cross-sectional surface in the SEM image of the TPU/BNNO@Co_3O_4@PPZ in Figure 4g. In addition, a few BNNO@Co_3O_4@PPZ exposed on the surface are tightly bonded with the TPU matrix in Figure 4h. These results indicate that the introduction of a PPZ layer can improve the interfacial interaction of the filler–matrix.

3.3. Thermal Properties of TPU Nanocomposites

The effects of different fillers on the pyrolysis behavior of TPU and its nanocomposites were tested by TG under nitrogen atmosphere, and the curves and data are displayed in Figure 5 and Table 1. The temperature at 5 wt% weight loss is defined as $Temp_{d5\%}$, while the temperature at maximum mass loss rate is named as $Temp_{max}$. According to the TGA curves in Figure 5a, except for TPU/BNNO@Co_3O_4, TPU/BNNO, and TPU/BNNO@Co_3O_4@PPZ, both suffer from the similar two-stage decomposition of TPU. In the first stage of decomposition, the degradation of the urethane bond in TPU produces diisocyanate, glycol, and carbon dioxide, while the second step is related to the thermal decomposition of the polyol segment in TPU soft segment [6]. As shown in Table 1, the introduction of Co_3O_4 greatly changed the pyrolysis behavior of TPU. The $Temp_{d5\%}$ value of TPU/BNNO@Co_3O_4 drops by 12.4 °C to 301.6 °C. According to Figure 5b, the addition of Co_3O_4 induces TPU degradation to a one-stage process, which is due to the catalytic degradation of TPU by transition metals, thus resulting in the early decomposition of the second stage [54]. However, the TPU/BNNO@Co_3O_4@PPZ showed the highest thermal decomposition temperature among other nanocomposites, and its $Temp_{d5\%}$ is 4.7 °C higher than TPU, which is attributed to the superior thermal stability and catalytic charring effect of PPZ. In addition, the char residue at 800 °C of TPU/BNNO@Co_3O_4 and TPU/BNNO@Co_3O_4@PPZ is higher than TPU. Due to the catalytic carbonization of PPZ and Co_3O_4, the char residue of TPU nanocomposites was increased to 7.8% and 7.42%, respectively. The formation of char residue can serve as a physical barrier that helps to inhibit the release of smoke and the transfer of mass and heat during the combustion, thus helping to improve the fire safety of TPU.

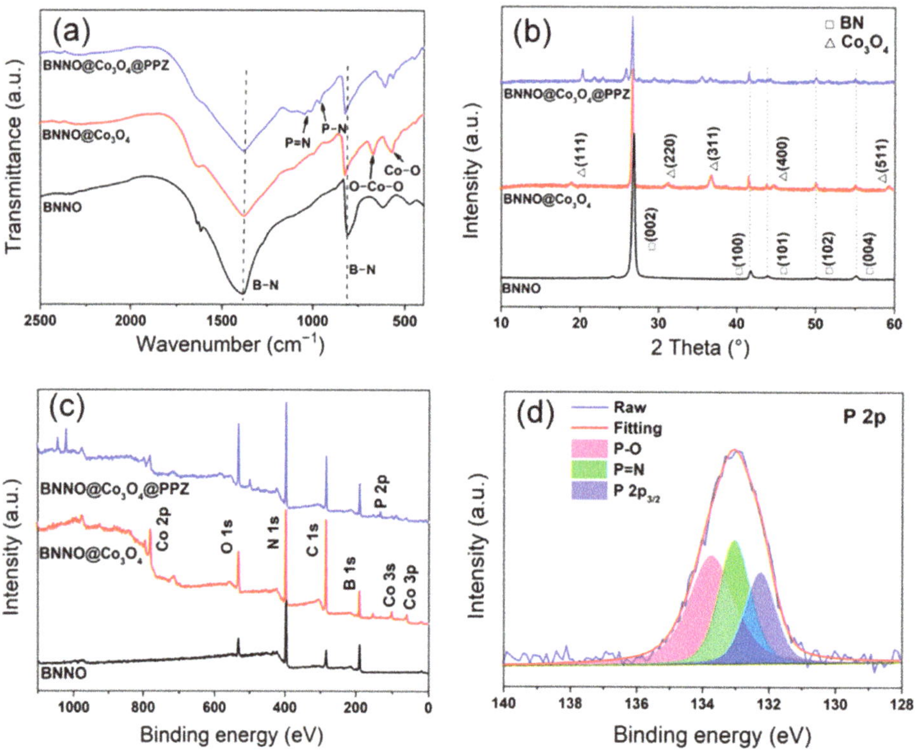

Figure 3. (**a**) FT-IR spectra; (**b**) XRD patterns; (**c**) XPS full survey of BNNO, BNNO@Co_3O_4, and BNNO@Co_3O_4@PPZ; and (**d**) the high resolution of P 2p of BNNO@Co_3O_4@PPZ.

Table 1. TGA data of TPU, TPU/BNNO, TPU/BNNO@Co_3O_4, and TPU/BNNO@Co_3O_4@PPZ nanocomposites.

Samples	$Temp_{d5\%}$ (°C)	$Temp_{max}$ (°C)	$R_{800°C}$ (wt%)
TPU	314.0	413.6	5.08
TPU/BNNO	314.2	413.8	6.81
TPU/BNNO@Co_3O_4	301.6	365.8	7.8
TPU/BNNO@Co_3O_4@PPZ	318.7	414.5	7.42

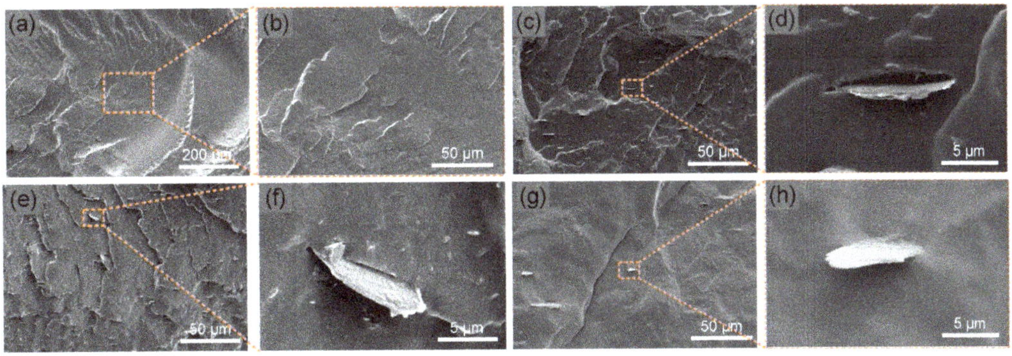

Figure 4. SEM images of the fractured surfaces of (**a**,**b**) TPU, (**c**,**d**) TPU/BNNO, (**e**,**f**) TPU/BNNO $@Co_3O_4$, and (**g**,**h**) TPU/BNNO$@Co_3O_4$@PPZ.

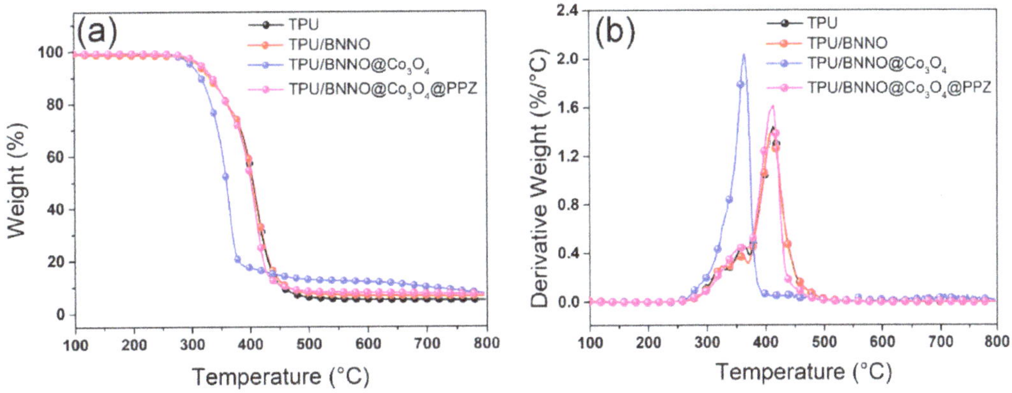

Figure 5. Thermal stability of TPU nanocomposites, (**a**) TGA curves, and (**b**) DTG curves at N_2 atmosphere.

3.4. Fire Safety of TPU Nanocomposites

Cone calorimetry can simulate the combustion state in the real environment and monitor the heat release and gas release during the combustion, which is a powerful means for assessing the fire performance of materials (Figure 6 and Table 2) [55]. In Figure 6a,b, it can be seen that an obvious suppression in PHRR and THR with the incorporation of BNNO$@Co_3O_4$ and BNNO$@Co_3O_4$@PPZ nanohybrid flame retardants. The BNNO$@Co_3O_4$@PPZ possesses the highest flame retardancy efficiency among all samples under the same filler content. With 2 wt% BNNO$@Co_3O_4$@PPZ, the PHRR and THR value of nanocomposites are reduced to 503.1 kW/m^2 and 56.9 MJ/m^2, respectively, which is 44.1% and 10.3% lower than that of pure TPU (900.8 kW/m^2 and 63.46 MJ/m^2). In Table 2, the TTI of pure TPU is 72 s, and with the appearance of Co_3O_4, the TTI of TPU/BNNO$@Co_3O_4$ and TPU/BNNO$@Co_3O_4$@PPZ reduced to 53 s and 68 s, respectively. This is ascribed to the degradation of TPU being accelerated due to catalytic degradation of Co_3O_4, which is consistent with TGA results [56,57].

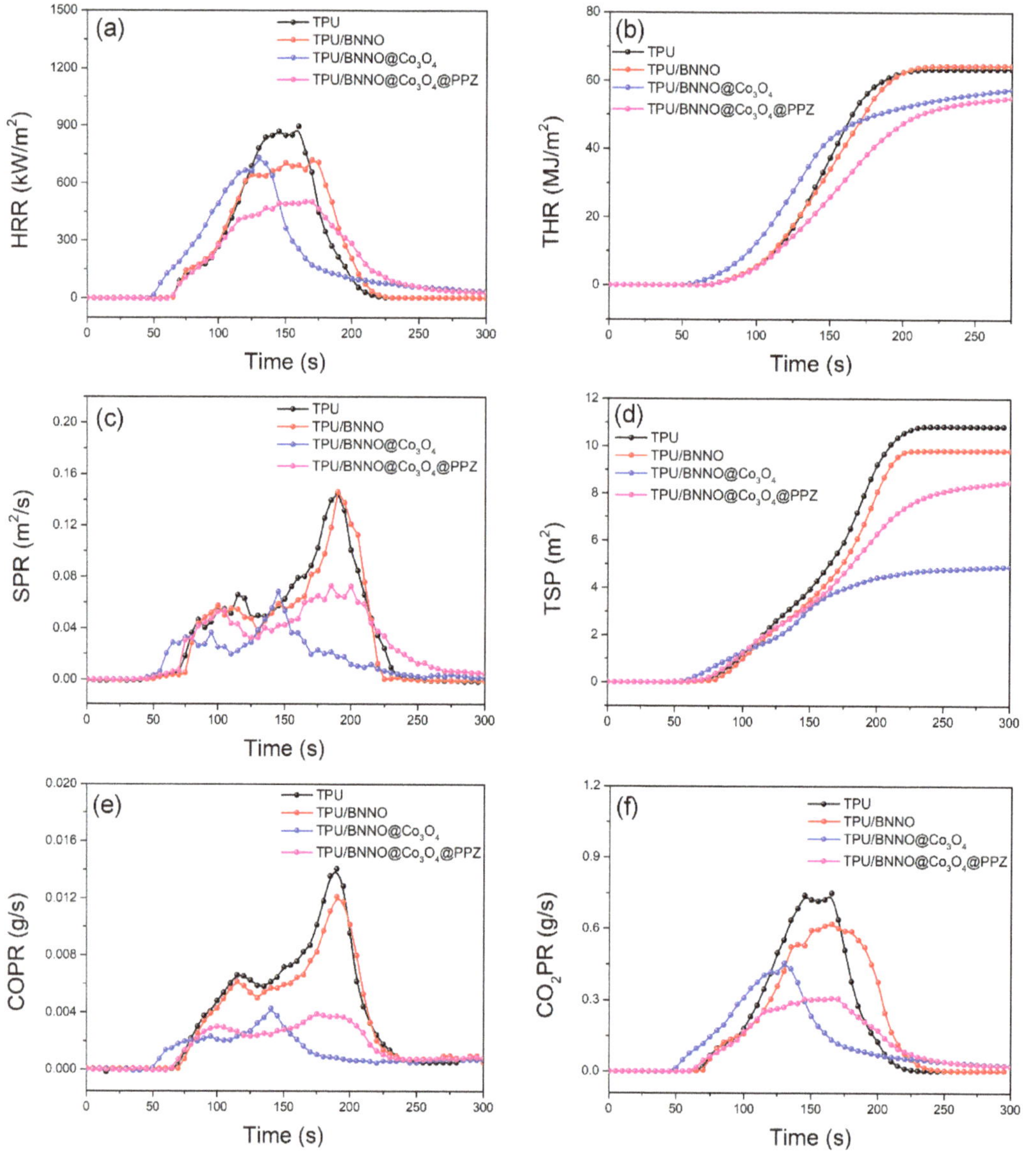

Figure 6. (**a**) HRR, (**b**) THR, (**c**) SPR, (**d**) TSR, (**e**) COPR, and (**f**) CO_2PR as a function of time of TPU and its nanocomposites obtained from the cone calorimeter.

Table 2. Relevant data of TPU nanocomposites by cone calorimeter test.

Samples	TTI(s)	PHRR (kW/m^2)	THR (MJ/m^2)	PSPR (m^2/s)	TSP (m^2)	PCOPR (g/s)	PCO$_2$PR (g/s)
Pure TPU	72	900.8	63.5	0.1497	10.8	0.0142	0.758
TPU/BNNO	72	724.7	64.5	0.1462	9.8	0.0123	0.663
TPU/BNNO@Co_3O_4	53	732.9	60.0	0.0685	4.9	0.0043	0.454
TPU/BNNO@Co_3O_4@PPZ	68	503.1	56.9	0.073	8.6	0.0039	0.308

Generally speaking, the spread of smoke is the biggest obstacle for humans escaping from a fire environment and the carbon monoxide (CO) from incomplete combustion directly threatens human life. Hence, the suppression effect on smoke is also a crucial index for assessing the performance of flame retardants. From Figure 6, the curves indicate that pure TPU produces huge amounts of smoke during combustion, with a PSPR of 0.1497 m^2/s and a high TSP value of 10.8 m^2. When BNNO is added, the PSPR and TSP of TPU/BNNO slightly decrease to 0.1462 m^2/s and 9.8 m^2, respectively. With 2 wt% BNNO@Co_3O_4 and BNNO@Co_3O_4@PPZ, the values of PSPR were decreased by 54.2% and 51.2%, respectively, exhibiting the superior efficiency of the nanohybrid flame retardants on smoke suppression. As for the release of CO, the PCOPR of TPU/BNNO@Co_3O_4 and TPU/BNNO@Co_3O_4@PPZ decrease from 69.7% and 72.5% to 0.0043 and 0.0039 g/s, respectively, much less than those to that of pure TPU (0.0142 g/s). Moreover, the peak values of carbon dioxide production (PCO_2PR) for TPU nanocomposites show a similar decline tendency as PCOPR. Compared to pure TPU, TPU/BNNO@Co_3O_4 and TPU/BNNO@Co_3O_4@PPZ show a 40.1 and 59.3% reduction in the PCO_2PR, respectively. The results of cone calorimetry show that the introduction of Co_3O_4 can effectively suppress toxic smoke production, especially for CO, during TPU combustion. The presence of PPZ in the hybrid flame retardant additive can further reduce the heat release and significantly enhance the fire performance due to the phosphorus element in PPZ [14,26,58].

3.5. Analysis of Char Residues after Combustion

To deeply explore the mechanism of the nanohybrid flame retardant, the digital photos and micromorphology of the char residues after the cone calorimeter test are shown in Figure 7. From the digital photos, loose and brittle residual char can be observed in pure TPU, and plentiful open holes are distributed on the surface. In contrast, the char formation of TPU nanocomposites containing flame retardants gradually becomes continuous and dense. According to the SEM images corresponding to the digital photos, for the pure TPU, the surface of the char residue is still honeycombed and densely covered with a large number of tiny holes on the microscopic scale, which provides access for flammable gases and toxic volatiles in and out during the combustion process, which is very unfavorable to fire safety. The introduction of BNNO and Co_3O_4 reduced the number of micropores, but the soft honeycomb-like structure can still be observed under high magnification. However, the char residue of TPU nanocomposites loaded with BNNO@Co_3O_4@PPZ nanohybrid flame retardant shows a compact and continuous structure, and there are tiny obvious holes in the field of vision. The formation of a dense char structure is ascribed to the catalytic charring effect of Co_3O_4 and PPZ during the combustion, which is conducive to retarding the mass and heat transfer and escape of pyrolysis volatiles, thus improving the flame retardancy of the TPU [59].

Generally speaking, the char-forming quality is positively correlated with its graphitization degree. The integrated area ratio of D-band and G-band (I_D/I_G) in Raman spectra is often used to reflect the graphitization degree of samples. The lower I_D/I_G means the higher graphitization degree and quality of the char layer [7,60]. In Figure 8, it is noteworthy that the values for I_D/I_G of pure TPU, TPU/BNNO, TPU/BNNO@Co_3O_4, and TPU/BNNO@Co_3O_4@PPZ are 3.06, 3.08, 2.98, and 2.85, respectively, which indicates the highest graphitization degree for the char residue of TPU/BNNO@Co_3O_4@PPZ. Furthermore, through the peak splitting of the high-resolution C1s XPS spectra, the concentration of atomic bond types of residual char can be quantitatively obtained to further assess the degree of graphitization. From Figure 9, the percentage of C-C bonds peak area of pure TPU, TPU/BNNO@Co_3O_4, and TPU/BNNO@Co_3O_4@PPZ are calculated by XPS analysis software AVANTAGE to be 51.48%, 65.28%, and 78.97%, respectively, which means the decreasing concentration of C-O, C=O, and oxidation degree. This result further demonstrates the higher graphitization degree of TPU/BNNO@Co_3O_4@PPZ.

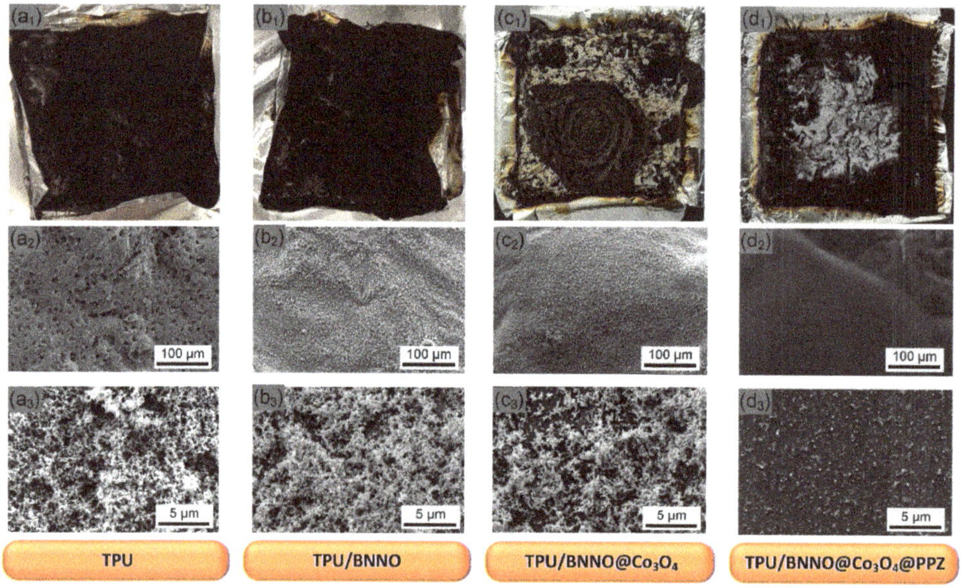

Figure 7. Digital photos of the char residues and SEM images (low magnification and high magnification) pure TPU (**a$_1$–a$_3$**), TPU/BNNO (**b$_1$–b$_3$**), TPU/BNNO@Co_3O_4 (**c$_1$–c$_3$**), and TPU/BNNO@Co_3O_4@PPZ (**d$_1$–d$_3$**).

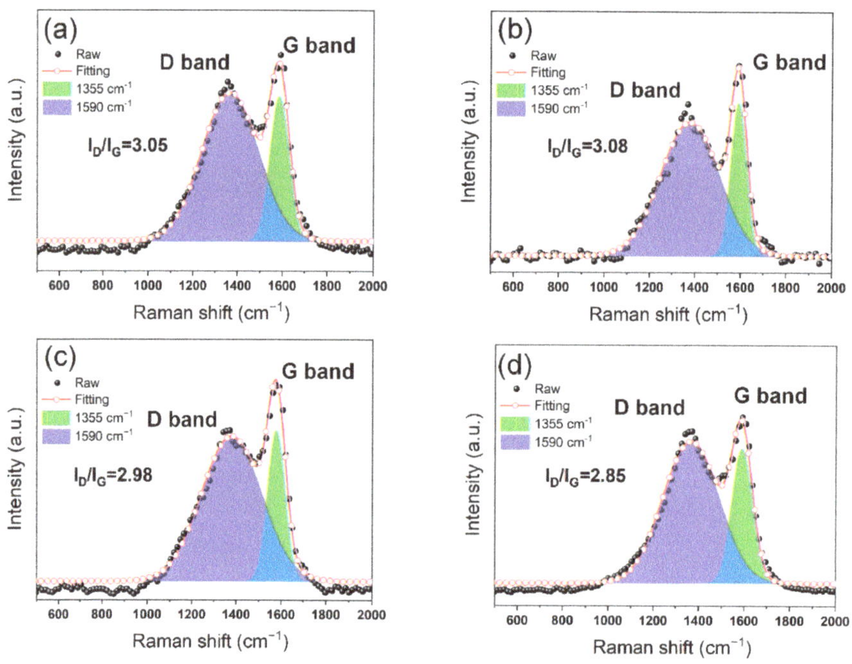

Figure 8. Raman spectra of char residue of (**a**) TPU, (**b**) TPU/BNNO, (**c**) TPU/BNNO@Co_3O_4, and (**d**) TPU/BNNO@Co_3O_4@PPZ.

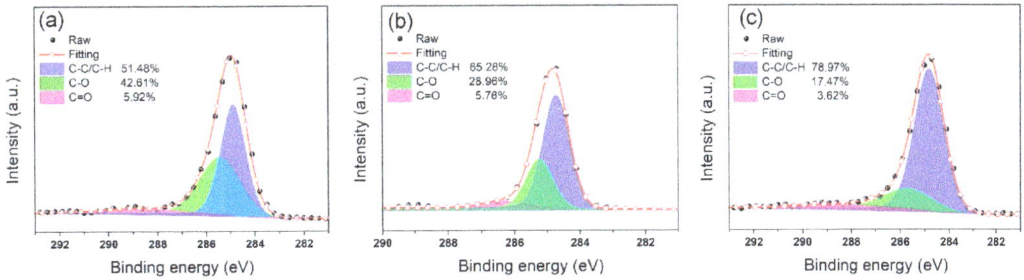

Figure 9. The high resolution of C1s XPS spectra of char residue of (**a**) pure TPU, (**b**) TPU/BNNO @Co_3O_4, and (c) TPU/BNNO@Co_3O_4@PPZ.

3.6. Vapor-Phase Analysis

The TG-IR technique was used to simulate the types and intensities of gases released during the pyrolysis of TPU and its nanocomposites under air atmosphere to further reveal the mechanism of the flame retardant improvement of TPU by BNNO@Co_3O_4 and BNNO@Co_3O_4@PPZ. The 3D TG-IR spectra depicted in Figure 10a–d demonstrate that the addition of Co_3O_4 significantly increases the intensity of gas volatilization, whereas the introduction of PPZ can suppress the release of volatile gaseous. In fact, most of the gas volatilization of TPU/BNNO@Co_3O_4 comes from the massive production of carbon dioxide (CO_2), which is of great significance for the production of non-combustible gases in flame retardant applications. Therefore, the release intensity and time of various gases must be analyzed in detail. According to Figure 10e, the intensity of CO released by TPU/BNNO@Co_3O_4 is the lowest at around 750–1250 s, and there is a strong but short period of CO release at around 1250 s. Meanwhile, the other three materials release a certain amount of CO in the time period of 750–1250 s, and a large amount of CO is released for a long time after 1500 s. Based on the CO_2 release curves in Figure, it can be concluded that the catalytic effect of transition metals in Co_3O_4 on decomposition makes CO, which is produced by incomplete combustion, and forming CO_2 when completely burned. The presence of the PPZ layer can suppress the release of CO due to its catalytic effect. However, the release of CO_2 is greatly weakened. This may be due to the coating of PPZ with strong thermal stability, and the catalytic effect of Co_3O_4 has not been fully exerted. In the real combustion process, PPZ is totally pyrolyzed at a higher temperature. Meanwhile, the flame retardant effects of Co_3O_4 and PPZ can be exerted at the same time. In addition, the release of carbonyl compounds and ethers is greatly suppressed during the combustion of TPU/BNNO@Co_3O_4@PPZ in Figure 10g,h, which is also due to the more complete combustion under the catalysis of Co_3O_4 and PPZ. The results of TG-FTIR can reasonably explain that BNNO shows the optimal inhibition effect on the release of gas degradation products of BNNO@Co_3O_4@PPZ due to the synergistic effects of Co_3O_4 and PPZ.

3.7. Flame Retardant Mechanism of TPU Nanocomposites

Based on the analysis of the gaseous and condensed phases, the excellent flame retardancy of BNNO@Co_3O_4@PPZ nanohybrid flame retardant, as shown in Figure 11, can be ascribed to the joint effect of BNNO, Co_3O_4, and PPZ. During the combustion, Co_3O_4 in the inner layer of nanohybrid flame retardant can catalyze CO, NO, and other combustible pyrolysis gases to generate CO_2, NO_2, and non-combustible gases, thus suppressing the toxic gases [41,56,61,62], which is consistent with the results of TG-IR. PPZ and Co_3O_4 can catalyze the formation of carbon, which is conducive to the formation of a denser char layer. The final dense char residue structure further restricted the permeation of oxygen and flammable products and acted as a barrier to heat. Meanwhile, the radicals, including •P, •PO, and •HPO generated during the pyrolysis of PPZ can combine with inflammable •H and •OH to remove flammable free radicals. This quenching effect would cut off the combustion and thus significantly reduce the heat release [12,43,63]. The "tortuous path" caused by the 2D layered structure of BNNO can also delay the transfer of combustible pyrolysis gas to the fire area to some extent. In addition, as a carrier, BNNO makes Co_3O_4 and PPZ disperse more uniformly and also enables toxic smoke to fully react with Co_3O_4 during escape.

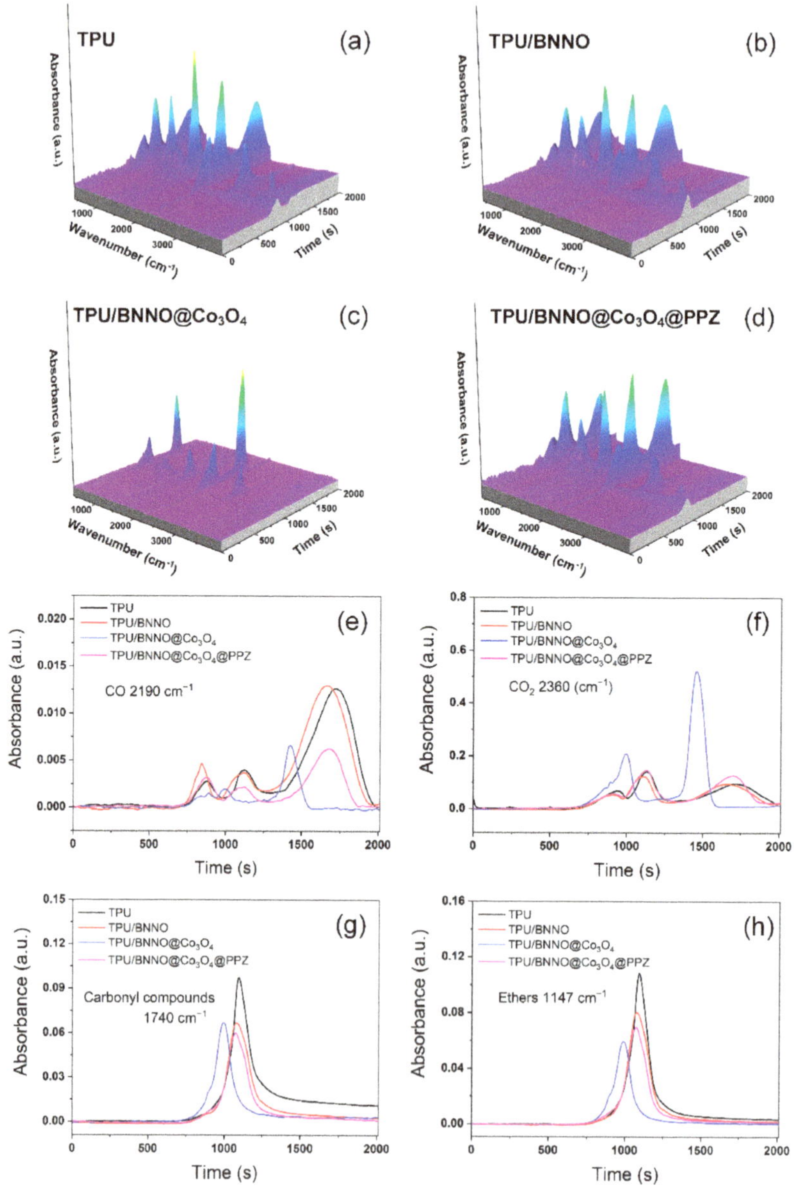

Figure 10. 3D TG-FTIR spectra of (**a**) TPU, (**b**) TPU/BNNO, (**c**) TPU/BNNO@Co_3O_4, and (**d**) TPU/BNNO@Co_3O_4@PPZ. The absorbance of pyrolysis products for TPU, TPU/BNNO, TPU/BNNO@Co_3O_4, and TPU/BNNO@Co_3O_4@PPZ as a function of time: (**e**) CO, (**f**) CO_2, (**g**) carbonyl compounds, and (**h**) ethers.

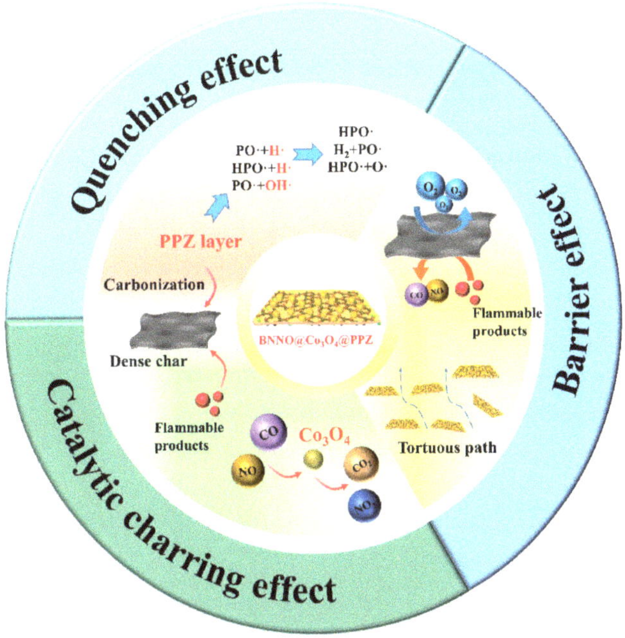

Figure 11. Diagram of flame-retardant mechanisms for TPU nanocomposites.

To highlight the significant improvement of BNNO@Co_3O_4@PPZ in the retardancy of TPU nanocomposites, a comprehensive comparison of many approaches devoted to suppressing the heat and toxic gas release during the TPU combustion, including this work, is given in Table 3. It can be seen that the direct introduction of cobalt-containing nanofillers [64] or the loading of cobalt on two-dimensional nanofillers [65] can only achieve the simultaneous suppression of heat release and smoke release with difficulty, and the thermal stability of the nanocomposite is deteriorated, which is not practical for specific applications. In this work, we hybridize cobalt (Co_3O_4) and phosphorus-containing polymers (PPZ) with BNNO. The introduction of BNNO@Co_3O_4@PPZ to TPU leads to remarkable reductions in PHRR (44.1%), PSPR (51.2%), PCOPR (72.5%), and PCO_2PR (59.3%), achieving simultaneous suppression of heat release and flue gas release, which are better than most reported results. Importantly, the retardancy of TPU is improved without deterioration of the thermal properties and mechanical properties.

Table 3. Comparisons in heat and toxic gases release of TPU nanocomposites reported in prior work and this work.

Sample	PHRR	PSPR	PCOPR	PCO_2PR
BN@P-PEI [22]	−34.4%	−26.8%	-	-
h-BN@SiO2@PA [24]	−23.5%	−29.2%	−26.8%	−11.0%
h-BN-PPy-PA-Cu2+ [25]	−35.6%	−31.8%	-	-
BNO/PANI [27]	−32.6%	+9.1%	0%	−32.2%
Co_3O_4-Tannic acid [56]	−18.9%	−33.3%	−11.4%	−10.5%
GO-DOPO [60]	−35.8%	−50%	−57.1%	−36.5%
Black phosphorus-HPL [64]	−49.9%	−45.8%	−37.5%	−32.5%
Co_3O_4/GNS [65]	−(<)10%	-	−18.2%	-
Co(OH)2 [66]	38.7%	−33.3%	−81.0%	-
BNNO@Co_3O_4@PPZ (This work)	−44.1%	−51.2%	−72.5%	−59.3%

3.8. Mechanical Properties of TPU Nanocomposites

The addition of flame retardant does not affect the excellent mechanical properties of TPU and can even be used as a reinforcing agent, which is the most ideal case. In Figure 12, good strength and toughness with a tensile strength of 33.58 MPa and an elongation at break of 2085% was obtained in pure TPU, while the existence of BNNO leads to a sharp decrease in the tensile properties of TPU/BNNO. The poor interfacial interaction, as shown in Figure 4c,d, between BNNO and TPU makes the sample easier to break under tensile stress. After the decoration of Co_3O_4, the tensile performance of TPU/BNNO@Co_3O_4 was enhanced compared to TPU/BNNO. It can be attributed that the introduction of Co_3O_4 nanoparticles can hinder the agglomeration of BNNO [65]. TPU/BNNO@Co_3O_4@PPZ obtained the highest tensile strength (34.77 MPa) and good elongation at break (2055%), basically maintaining the original mechanical properties of TPU. This is mainly because the organic PPZ layer improves the interfacial adhesion between the filler–matrix, thus realizing the enhancement of mechanical properties of TPU@Co_3O_4@PPZ [41].

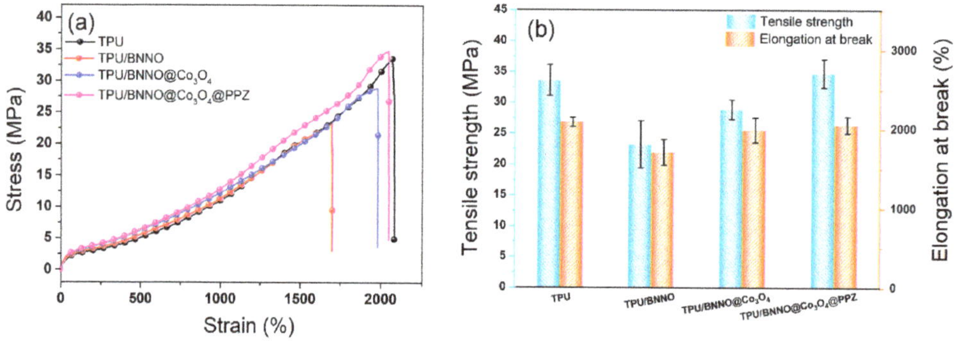

Figure 12. (**a**) Strain–stress curves, (**b**) tensile properties of TPU nanocomposites.

4. Conclusions

In this work, the transition metal oxide Co_3O_4 was decorated onto the surface of oxidized h-BN (BNNO@Co_3O_4). Then the PPZ was coated on the surface of BNNO@Co_3O_4 (BNNO@Co_3O_4@PPZ) via the high-temperature polymerization of HCCP. With the content of 2 wt% BNNO@Co_3O_4@PPZ, the PHRR and THR of the TPU/BNNO@Co_3O_4@PPZ nanocomposite were significantly reduced by 44.1% and 10.4%, respectively, as compared with those of pure TPU. Moreover, BNNO@Co_3O_4@PPZ also exhibited a remarkable suppression of smoke production. The PSPR, the PCOR, and the PCO_2R of TPU/BNNO @Co_3O_4@PPZ nanocomposites had a substantial decline of 51.2, 72.5, and 59.4%, respectively. After analyzing the gaseous and condensed phases of combustion products of nanocomposites, the main mechanisms for the nanohybrid flame retardant BNNO@Co_3O_4 @PPZ to enhance the safety performance of TPU during combustion were ascribed to the quenching effect of pyrolytic products of PPZ on flammable-free radicals and the barrier effect of the dense and continuous graphitized char layer formed by the catalytic charring effect of PPZ and Co_3O_4 on heat and gas delivery. Moreover, SEM graphs showed that BNNO@Co_3O_4@PPZ was uniformly dispersed in the TPU matrix due to the coating of PPZ and exhibited a close interface bonding between filler and polymer matrix. Thus, the BNNO@Co_3O_4@PPZ was able to improve the fire safety of TPU while maintaining its original mechanical properties. This work provides a simple and effective method and structure of a hybrid flame retardant for improving the flame retardancy and smoke suppression of TPU during ignition.

Supplementary Materials: The following supporting information can be downloaded at: https://www.mdpi.com/article/10.3390/polym14204341/s1, Figure S1: (a) TEM image, (b) EDX elemental mapping of B, N, P, and Co of BNNO@Co_3O_4@PPZ. Figure S2: High resolution of Co 2p of BNNO@Co_3O_4@PPZ; Figure S3: TGA curves of BNNO, BNNO@Co_3O_4, and BNNO@Co_3O_4@PPZ.

Author Contributions: Conceptualization, Y.T., W.W., W.Z. and X.C.; methodology, Y.T., W.W., W.Z. and X.Y.; validation, Y.X. and H.Z.; investigation, W.Z. and B.Y.; resources, W.W., C.W., T.B.Y.C. and A.C.Y.Y.; data curation, A.C.Y.Y.; writing—original draft preparation, Y.T.; writing—review and editing, W.W., X.C. and X.Y.; supervision, X.C. and X.Y.; project administration, X.C. and X.Y.; funding acquisition, X.C., W.W. and A.C.Y.Y. All authors have read and agreed to the published version of the manuscript.

Funding: This research was funded by the Guangdong Basic and Applied Basic Research Foundation (2021A1515012425), the National Natural Science Foundation of China (52103029), the International Collaboration Programs of Guangdong Province (2020A0505100010), the Opening Project of Key Laboratory of Polymer Processing Engineering (South China University of Technology), Ministry of Education of China (KFKT1904), Australian Research Council Industrial Training Transformation Centre (IC170100032) and Guangzhou Innovation Leading Team Project (201809010011).

Institutional Review Board Statement: Not applicable.

Informed Consent Statement: Not applicable.

Data Availability Statement: The raw/processed data generated in this work are available upon request from the corresponding author.

Conflicts of Interest: The authors declare no conflict of interest.

References

1. Kim, S.; Jeon, H.; Shin, S.; Park, S.; Park, J. Superior toughness and fast self-healing at room temperature engineered by transparent elastomers. *Adv. Mater.* **2017**, *30*, 1705145. [CrossRef] [PubMed]
2. Chen, Q.; Gao, Q.; Wang, X.; Schubert, D.W.; Liu, X. Flexible, conductive, and anisotropic thermoplastic polyurethane/polydopamine/MXene foam for piezoresistive sensors and motion monitoring. *Compos. Part A Appl. Sci. Manuf.* **2022**, *155*, 106838. [CrossRef]
3. Chen, T.; Xie, Y.; Wang, Z.; Lou, J.; Liu, D.; Xu, R.; Cui, Z.; Li, S.; Panahi-Sarmad, M.; Xiao, X. Recent advances of flexible strain sensors based on conductive fillers and thermoplastic polyurethane matrixes. *ACS Appl. Polym. Mater.* **2021**, *3*, 5317–5338. [CrossRef]
4. Zhao, H.; Gao, W.; Li, Q.; Khan, M.R.; Hu, G.H.; Liu, Y.; Wu, W.; Huang, C.X.; Li, R.K.Y. Recent advances in superhydrophobic polyurethane: Preparations and applications. *Adv. Colloid Interface Sci.* **2022**, *303*, 102644. [CrossRef] [PubMed]
5. Jamsaz, A.; Goharshadi, E.K.; Barras, A.; Ifires, M.; Szunerits, S.; Boukherroub, R. Magnetically driven superhydrophobic/superoleophilic graphene-based polyurethane sponge for highly efficient oil/water separation and demulsification. *Sep. Purif. Technol.* **2021**, *274*, 118931. [CrossRef]
6. Wan, L.; Deng, C.; Chen, H.; Zhao, Z.-Y.; Huang, S.-C.; Wei, W.-C.; Yang, A.-H.; Zhao, H.-B.; Wang, Y.-Z. Flame-retarded thermoplastic polyurethane elastomer: From organic materials to nanocomposites and new prospects. *Chem. Eng. J.* **2021**, *417*, 129314. [CrossRef]
7. Wu, W.; Zhao, W.; Gong, X.; Sun, Q.; Cao, X.; Su, Y.; Yu, B.; Li, R.K.Y.; Vellaisamy, R.A.L. Surface decoration of Halloysite nanotubes with POSS for fire-safe thermoplastic polyurethane nanocomposites. *J. Mater. Sci. Technol.* **2021**, *101*, 107–117. [CrossRef]
8. Jiao, C.; Li, M.; Chen, X.; Li, S. Flame retardancy and thermal decomposition behavior of TPU/chitosan composites. *Polym. Adv. Technol.* **2019**, *31*, 178–188. [CrossRef]
9. Liu, Q.; Wang, H.; Li, H.; Sun, J.; Gu, X.; Zhang, S. Constructing a novel synergistic flame retardant by hybridization of zeolitic imidazolate framework-67 and graphene oxide for thermoplastic polyurethane. *Polym. Adv. Technol.* **2022**, *33*, 2374–2385. [CrossRef]
10. Jamsaz, A.; Goharshadi, E.K. Flame retardant, superhydrophobic, and superoleophilic reduced graphene oxide/orthoaminophenol polyurethane sponge for efficient oil/water separation. *J. Mol. Liq.* **2020**, *307*, 112979. [CrossRef]
11. Yu, B.; Tawiah, B.; Wang, L.-Q.; Yuen, A.C.Y.; Zhang, Z.-C.; Shen, L.-L.; Lin, B.; Fei, B.; Yang, W.; Li, A.; et al. Interface decoration of exfoliated MXene ultra-thin nanosheets for fire and smoke suppressions of thermoplastic polyurethane elastomer. *J. Hazard. Mater.* **2019**, *374*, 110–119. [CrossRef] [PubMed]
12. Huang, W.; Huang, J.; Yu, B.; Meng, Y.; Cao, X.; Zhang, Q.; Wu, W.; Shi, D.; Jiang, T.; Li, R.K. Facile preparation of phosphorus containing hyperbranched polysiloxane grafted graphene oxide hybrid toward simultaneously enhanced flame retardancy and smoke suppression of thermoplastic polyurethane nanocomposites. *Compos. Part A Appl. Sci. Manuf.* **2021**, *150*, 106614. [CrossRef]

13. Cai, W.; Zhan, J.; Feng, X.; Yuan, B.; Liu, J.; Hu, W.; Hu, Y. Facile Construction of Flame-Retardant-Wrapped Molybdenum Disulfide Nanosheets for Properties Enhancement of Thermoplastic Polyurethane. *Ind. Eng. Chem. Res.* **2017**, *56*, 7229–7238. [CrossRef]
14. Huang, S.-C.; Deng, C.; Zhao, Z.-Y.; Chen, H.; Gao, Y.-Y.; Wang, Y.-Z. Phosphorus-containing organic-inorganic hybrid nanoparticles for the smoke suppression and flame retardancy of thermoplastic polyurethane. *Polym. Degrad. Stab.* **2020**, *178*, 109179. [CrossRef]
15. Qian, Y.; Su, W.; Li, L.; Fu, H.; Li, J.; Zhang, Y. Synthesis of 3D hollow layered double hydroxide-molybdenum disulfide hybrid materials and their application in flame retardant thermoplastic polyurethane. *Polymers* **2022**, *14*, 1506. [CrossRef]
16. Huang, S.-C.; Deng, C.; Chen, H.; Li, Y.-M.; Zhao, Z.-Y.; Wang, S.-X.; Wang, Y.-Z. Novel ultrathin layered double hydroxide nanosheets with in situ formed oxidized phosphorus as anions for simultaneous fire resistance and mechanical enhancement of thermoplastic polyurethane. *ACS Appl. Polym. Mater.* **2019**, *1*, 1979–1990. [CrossRef]
17. Mokoena, T.E.; Magagula, S.I.; Mochane, M.J.; Mokhena, T.C. Mechanical properties, thermal conductivity, and modeling of boron nitride-based polymer composites: A review. *Express Polym. Lett.* **2021**, *15*, 1148–1173. [CrossRef]
18. Zheng, Z.; Cox, M.; Li, B. Surface modification of hexagonal boron nitride nanomaterials: A review. *J. Mater. Sci.* **2018**, *53*, 66–99. [CrossRef]
19. Fischer, A.J.; Zhong, Y.; Zhang, L.; Wu, W.; Drummer, D. Heat propagation in thermally conductive polymers of PA6 and hexagonal boron nitride. *Fire Mater.* **2019**, *43*, 928–935. [CrossRef]
20. Yin, L.; Gong, K.; Zhou, K.; Qian, X.; Shi, C.; Gui, Z.; Qian, L. Flame-retardant activity of ternary integrated modified boron nitride nanosheets to epoxy resin. *J. Colloid Interface Sci.* **2021**, *608*, 853–863. [CrossRef]
21. Zhang, S.; Sun, H.; Lan, T.; Xue, X.; Liu, X. Polybenzoxazine/boron nitride foam: A promising low-k, flame-retardant and robust material. *J. Mater. Sci.* **2021**, *56*, 18749–18761. [CrossRef]
22. Cai, W.; Hong, N.; Feng, X.; Zeng, W.; Shi, Y.; Zhang, Y.; Wang, B.; Hu, Y. A facile strategy to simultaneously exfoliate and functionalize boron nitride nanosheets via Lewis acid-base interaction. *Chem. Eng. J.* **2017**, *330*, 309–321. [CrossRef]
23. Cai, W.; Mu, X.; Pan, Y.; Guo, W.; Wang, J.; Yuan, B.; Feng, X.; Tai, Q.; Hu, Y. Facile fabrication of organically modified boron nitride nanosheets and its effect on the thermal stability, flame retardant, and mechanical properties of thermoplastic polyurethane. *Polym. Adv. Technol.* **2018**, *29*, 2545–2552. [CrossRef]
24. Cai, W.; Wang, B.; Liu, L.; Zhou, X.; Chu, F.; Zhan, J.; Hu, Y.; Kan, Y.; Wang, X. An operable platform towards functionalization of chemically inert boron nitride nanosheets for flame retardancy and toxic gas suppression of thermoplastic polyurethane. *Compos. Part B Eng.* **2019**, *178*, 107462. [CrossRef]
25. Wang, J.; Zhang, D.; Zhang, Y.; Cai, W.; Yao, C.; Hu, Y.; Hu, W. Construction of multifunctional boron nitride nanosheet towards reducing toxic volatiles (CO and HCN) generation and fire hazard of thermoplastic polyurethane. *J. Hazard. Mater.* **2019**, *362*, 482–494. [CrossRef]
26. Yin, S.H.; Ren, X.L.; Lian, P.C.; Zhu, Y.Z.; Mei, Y. Synergistic Effects of Black Phosphorus/Boron Nitride Nanosheets on Enhancing the Flame-Retardant Properties of Waterborne Polyurethane and Its Flame-Retardant Mechanism. *Polymers* **2020**, *12*, 1487. [CrossRef]
27. Zhi, Y.-R.; Yu, B.; Yuen, A.C.Y.; Liang, J.; Wang, L.-Q.; Yang, W.; Lu, H.-D.; Yeoh, G. Surface Manipulation of Thermal-Exfoliated Hexagonal Boron Nitride with Polyaniline for Improving Thermal Stability and Fire Safety Performance of Polymeric Materials. *ACS Omega* **2018**, *3*, 14942–14952. [CrossRef]
28. Wang, H.; Qiao, H.; Guo, J.; Sun, J.; Li, H.; Zhang, S.; Gu, X. Preparation of cobalt-based metal organic framework and its application as synergistic flame retardant in thermoplastic polyurethane (TPU). *Compos. Part B Eng.* **2020**, *182*, 107498. [CrossRef]
29. Zhou, K.; Gong, K.; Gao, F.; Yin, L. Facile strategy to synthesize MXene@LDH nanohybrids for boosting the flame retardancy and smoke suppression properties of epoxy. *Compos. Part A Appl. Sci. Manuf.* **2022**, *157*, 106912. [CrossRef]
30. Gong, K.; Yin, L.; Zhou, K.; Qian, X.; Shi, C.; Gui, Z.; Yu, B.; Qian, L. Construction of interface-engineered two-dimensional nanohybrids towards superb fire resistance of epoxy composites. *Compos. Part A Appl. Sci. Manuf.* **2022**, *152*, 106707. [CrossRef]
31. Kalali, E.N.; Guo, W.; Wang, X.; Xing, W.; Song, L.; Hu, Y. Effect of metal-based nanoparticles decorated graphene hybrids on flammability of epoxy nanocomposites. *Compos. Part A Appl. Sci. Manuf.* **2020**, *129*, 105694. [CrossRef]
32. Zhu, W.; Gao, X.; Li, Q.; Li, H.; Chao, Y.; Li, M.; Mahurin, S.M.; Li, H.; Zhu, H.; Dai, S. Controlled Gas Exfoliation of Boron Nitride into Few-Layered Nanosheets. *Angew. Chem. Int. Ed.* **2016**, *55*, 10766–10770. [CrossRef] [PubMed]
33. Chen, S.; Xu, R.; Liu, J.; Zou, X.; Qiu, L.; Kang, F.; Liu, B.; Cheng, H.-M. Simultaneous Production and Functionalization of Boron Nitride Nanosheets by Sugar-Assisted Mechanochemical Exfoliation. *Adv. Mater.* **2019**, *31*, e1804810. [CrossRef] [PubMed]
34. Liu, H.; Du, Y.; Lei, S.; Liu, Z. Flame-retardant activity of modified boron nitride nanosheets to cotton. *Text. Res. J.* **2019**, *90*, 512–522. [CrossRef]
35. Yu, B.; Xing, W.; Guo, W.; Qiu, S.; Wang, X.; Lo, S.; Hu, Y. Thermal exfoliation of hexagonal boron nitride for effective enhancements on thermal stability, flame retardancy and smoke suppression of epoxy resin nanocomposites via sol–gel process. *J. Mater. Chem. A* **2016**, *4*, 7330–7340. [CrossRef]
36. Wang, D.; Mu, X.; Cai, W.; Song, L.; Ma, C.; Hu, Y. Constructing phosphorus, nitrogen, silicon-co-contained boron nitride nanosheets to reinforce flame retardant properties of unsaturated polyester resin. *Compos. Part A Appl. Sci. Manuf.* **2018**, *109*, 546–554. [CrossRef]

37. Yang, L.; Guo, J.; Zhang, L.; Li, C. Significant Improvement in the Flame Retardancy and Thermal Conductivity of the Epoxy Resin via Constructing a Branched Flame Retardant Based on SI-ATRP Initiated by Dopamine-Modified Boron Nitride. *Ind. Eng. Chem. Res.* **2022**, *61*, 8031–8042. [CrossRef]
38. Zhou, X.; Qiu, S.; Mu, X.; Zhou, M.; Cai, W.; Song, L.; Xing, W.; Hu, Y. Polyphosphazenes-based flame retardants: A review. *Compos. Part B Eng.* **2020**, *202*, 108397. [CrossRef]
39. Singh, K.P.; Mishra, A.; Kumar, N.; Tripathi, D.; Shami, T.C. Evaluation of thermal, morphological and flame-retardant properties of thermoplastic polyurethane/polyphosphazene blends. *Polym. Bull.* **2018**, *75*, 2415–2430. [CrossRef]
40. Qiu, S.; Zhou, Y.; Zhou, X.; Zhang, T.; Wang, C.; Yuen, K.K.R.; Hu, W.; Hu, Y. Air-Stable Polyphosphazene-Functionalized Few-Layer Black Phosphorene for Flame Retardancy of Epoxy Resins. *Small* **2019**, *15*, e1805175. [CrossRef]
41. Qiu, S.; Hu, Y.; Shi, Y.; Hou, Y.; Kan, Y.; Chu, F.; Sheng, H.; Yuen, K.K.R.; Xing, W. In situ growth of polyphosphazene particles on molybdenum disulfide nanosheets for flame retardant and friction application. *Compos. Part A Appl. Sci. Manuf.* **2018**, *114*, 407–417. [CrossRef]
42. Guo, Y.; Zhang, S.; Wang, G.; Zhu, Y. Fabrication of Anisotropic Polyphosphazene/Bio-based Poly(urethane-acrylate) composite foams with High Thermal Insulation and Flame Retardancy. *Polymer* **2021**, *231*, 124108. [CrossRef]
43. Hong, J.; Wu, T.; Wang, X.; Lu, Z.; Zhang, J.; Zeng, B.; Yuan, C.; Dai, L. Copper-catalyzed pyrolysis of halloysites@polyphosphazene for efficient carbonization and smoke suppression. *Compos. Part B Eng.* **2022**, *230*, 109547. [CrossRef]
44. Qiu, S.; Xing, W.; Mu, X.; Feng, X.; Ma, C.; Yuen, K.K.R.; Hu, Y. A 3D Nanostructure Based on Transition-Metal Phosphide Decorated Heteroatom-Doped Mesoporous Nanospheres Interconnected with Graphene: Synthesis and Applications. *ACS Appl. Mater. Interfaces* **2016**, *8*, 32528–32540. [CrossRef] [PubMed]
45. Yu, H.; Cui, J.; Zhang, H.; Yang, B.; Guo, J.; Mu, B.; Wang, Z.; Li, H.; Tian, L. A novel flame retardant consisting of functionalized Salen-Ni based polyphosphazene microspheres. *High Perform. Polym.* **2022**, *34*, 914–927. [CrossRef]
46. Zhou, X.; Qiu, S.; Xing, W.; Gangireddy, C.S.R.; Gui, Z.; Hu, Y. Hierarchical Polyphosphazene@Molybdenum Disulfide Hybrid Structure for Enhancing the Flame Retardancy and Mechanical Property of Epoxy Resins. *ACS Appl. Mater. Interfaces* **2017**, *9*, 29147–29156. [CrossRef]
47. Wang, X.; Xing, W.; Feng, X.; Yu, B.; Lu, H.; Song, L.; Hu, Y. The effect of metal oxide decorated graphene hybrids on the improved thermal stability and the reduced smoke toxicity in epoxy resins. *Chem. Eng. J.* **2014**, *250*, 214–221. [CrossRef]
48. Guender, D.; Watanabe, K.; Taniguchi, T.; Witte, G. Van der Waals Bound Organic/2D Insulator Hybrid Structures: Epitaxial Growth of Acene Films on *h*BN(001) and the Influence of Surface Defects. *ACS Appl. Mater. Interfaces* **2020**, *12*, 38757–38767. [CrossRef]
49. Mu, X.; Pan, Y.; Ma, C.; Zhan, J.; Song, L. Novel Co_3O_4/covalent organic frameworks nanohybrids for conferring enhanced flame retardancy, smoke and CO suppression and thermal stability to polypropylene. *Mater. Chem. Phys.* **2018**, *215*, 20–30. [CrossRef]
50. Ponnamma, D.; Nair, S.S.; Parangusan, H.; Hassan, M.K.; Adham, S.; Karim, A.; Al-Maadeed, M.A.A. White Graphene-Cobalt Oxide Hybrid Filler Reinforced Polystyrene Nanofibers for Selective Oil Absorption. *Polymers* **2019**, *12*, 4. [CrossRef]
51. Zhou, S.; Wang, G.; Xie, Y.; Wang, H.; Bai, J. Synthesis of carbon-coated Co_3O_4 composite with dendrite-like morphology and its electrochemical performance for lithium-ion batteries. *J. Nanoparticle Res.* **2013**, *15*, 1740. [CrossRef]
52. Wang, Y.; Wang, C.; Wang, Y.; Liu, H.; Huang, Z. Superior Sodium-Ion Storage Performance of Co_3O_4@Nitrogen-Doped Carbon: Derived from a Metal-Organic Framework. *J. Mater. Chem. A* **2016**, *4*, 5428–5435. [CrossRef]
53. Tong, Y.-Z.; Zhao, W.-J.; Wu, W.; Zhang, D.-L.; He, G.-J.; Yang, Z.-T.; Cao, X.-W. Realizing enhanced dielectric and mechanical performance of polyvinylidene fluoride/SiC nanocomposites through a bio-inspired interface design. *Adv. Compos. Hybrid Mater.* **2022**, *5*, 263–277. [CrossRef]
54. Nabipour, H.; Wang, X.; Song, L.; Hu, Y. Metal-organic frameworks for flame retardant polymers application: A critical review. *Compos. Part A Appl. Sci. Manuf.* **2020**, *139*, 106113. [CrossRef]
55. Quan, Y.; Zhang, Z.; Tanchak, R.N.; Wang, Q. A review on cone calorimeter for assessment of flame-retarded polymer composites. *J. Therm. Anal. Calorim.* **2022**, *147*, 10209–10234. [CrossRef]
56. Wang, X.; Cai, W.; Ye, D.; Zhu, Y.; Cui, M.; Xi, J.; Liu, J.; Xing, W. Bio-based polyphenol tannic acid as universal linker between metal oxide nanoparticles and thermoplastic polyurethane to enhance flame retardancy and mechanical properties. *Compos. Part B Eng.* **2021**, *224*, 109206. [CrossRef]
57. Li, Z.; Wang, D.-Y. Nano-architectured mesoporous silica decorated with ultrafine Co_3O_4 toward an efficient way to delaying ignition and improving fire retardancy of polystyrene. *Mater. Des.* **2017**, *129*, 69–81. [CrossRef]
58. Liu, C.; Yao, A.; Chen, K.; Shi, Y.; Feng, Y.; Zhang, P.; Yang, F.; Liu, M.; Chen, Z. MXene based core-shell flame retardant towards reducing fire hazards of thermoplastic polyurethane. *Compos. Part B Eng.* **2021**, *226*, 109363. [CrossRef]
59. Li, H.; Ning, N.; Zhang, L.; Wang, Y.; Liang, W.; Tian, M.; Chan, T.W. Effect of content of organophosphorus on flame retardancy mode of thermoplastic polyurethane. *Polymer* **2015**, *67*, 1–11. [CrossRef]
60. Cao, X.; Zhao, W.; Huang, J.; He, Y.; Liang, X.; Su, Y.; Wu, W.; Li, R.K. Interface engineering of graphene oxide containing phosphorus/nitrogen towards fire safety enhancement for thermoplastic polyurethane. *Compos. Commun.* **2021**, *27*, 100821. [CrossRef]
61. Pan, H.; Ma, W.; Zhang, Z.; Liu, Y.; Lu, F.; Yu, B.; Zhang, X. Co-Effect Flame Retardation of Co_3O_4-Loaded Titania Nanotubes and α-Zirconium Phosphate in the Epoxy Matrix. *ACS Omega* **2020**, *5*, 28475–28482. [CrossRef] [PubMed]

62. Feng, Y.; He, C.; Wen, Y.; Ye, Y.; Zhou, X.; Xie, X.; Mai, Y.-W. Improving thermal and flame retardant properties of epoxy resin by functionalized graphene containing phosphorous, nitrogen and silicon elements. *Compos. Part A Appl. Sci. Manuf.* **2017**, *103*, 74–83. [CrossRef]
63. Wu, W.; He, H.; Liu, T.; Wei, R.; Cao, X.; Sun, Q.; Venkatesh, S.; Yuen, R.K.; Roy, V.A.; Li, R.K. Synergetic enhancement on flame retardancy by melamine phosphate modified lignin in rice husk ash filled P34HB biocomposites. *Compos. Sci. Technol.* **2018**, *168*, 246–254. [CrossRef]
64. Qiu, S.; Liang, J.; Hou, Y.; Zhou, X.; Zhou, Y.; Wang, J.; Zou, B.; Xing, W.; Hu, Y. Hindered phenolic antioxidant passivation of black phosphorus affords air stability and free radical quenching. *J. Colloid Interface Sci.* **2022**, *606*, 1395–1409. [CrossRef]
65. Zhou, K.; Gui, Z.; Hu, Y.; Jiang, S.; Tang, G. The influence of cobalt oxide–graphene hybrids on thermal degradation, fire hazards and mechanical properties of thermoplastic polyurethane composites. *Compos. Part A Appl. Sci. Manuf.* **2016**, *88*, 10–18. [CrossRef]
66. Zhang, J.; Kong, Q.; Yang, L.; Wang, D.-Y. Few layered $Co(OH)_2$ ultrathin nanosheet-based polyurethane nanocomposites with reduced fire hazard: From eco-friendly flame retardance to sustainable recycling. *Green Chem.* **2016**, *18*, 3066–3074. [CrossRef]

MDPI
St. Alban-Anlage 66
4052 Basel
Switzerland
www.mdpi.com

Polymers Editorial Office
E-mail: polymers@mdpi.com
www.mdpi.com/journal/polymers

www.ingramcontent.com/pod-product-compliance
Lightning Source LLC
Chambersburg PA
CBHW061620210926
51850CB00009B/2384
* 9 7 8 3 7 2 5 8 0 3 3 7 8 *